APPLICATION OF CATHODOLL
TO THE STUDY OF SEDIMENTARY ROCKS

Minerals in sedimentary rocks, such as quartz, feldspar, and carbonate minerals, emit characteristic visible luminescence called cathodoluminescence (CL) when bombarded by high-energy electrons in a suitable instrument. CL emissions can be captured and displayed as color images in a cathodoluminescence microscope or as high-resolution monochromatic images in a scanning electron microscope. CL imaging is particularly useful for studying sedimentary rocks because it provides information, not readily available by other techniques, about the provenance of the mineral grains that constitute sedimentary rocks. CL images also provide insights, not available by other research techniques, into diagenetic changes, such as cementation and porosity loss, which take place in sandstones, shales, and carbonate rocks during burial.

The book begins with an easily understood presentation of the fundamental principles of CL imaging. This presentation is followed by a description and discussion of the instruments used to obtain CL images. Finally, the principal applications of CL imaging to study of sedimentary rocks are described in detail. This short guide provides the first comprehensive, focused, easily understood description of the various applications of cathodoluminescence imaging to study of sedimentary rocks. It will be an important resource for academic researchers, industry professionals and advanced graduate students in sedimentary geology.

SAM BOGGS received his Ph.D. degree from the University of Colorado and worked in the petroleum industry for a number of years before beginning an academic career. After joining the University of Oregon he taught courses and carried out research in sedimentary petrology, stratigraphy, field geology, petroleum geology, and geological oceanography. He continues to do research as a professor emeritus at the University of Oregon and is a senior fellow of the Geological Association of America.

DAVID KRINSLEY is a courtesy professor at the University of Oregon, and a professor emeritus at Arizona State University. His current research interests include the study of both coarse- and fine-grained sediments using scanning and transmission electron microscopy. His particular interests are the examination of sandstones and shales with scanning electron microscopy combined with time-of-flight scanning mass spectrometry. Professor Krinsley is a fellow of the American Association for the Advancement of Science and a senior fellow of the Geological Association of America.

Sam Boggs and David Krinsley have published numerous other books including *Backscattered Scanning Electron Microscopy and Image Analysis of Sediments and Sedimentary Rocks* with co-authors Kenneth Pye and Keith Tovey (Cambridge, 1998).

APPLICATION OF CATHODOLUMINESCENCE IMAGING TO THE STUDY OF SEDIMENTARY ROCKS

SAM BOGGS, JR. AND DAVID KRINSLEY
University of Oregon

CAMBRIDGE UNIVERSITY PRESS
Cambridge, New York, Melbourne, Madrid, Cape Town, Singapore,
São Paulo, Delhi, Dubai, Tokyo, Mexico City

Cambridge University Press
The Edinburgh Building, Cambridge CB2 8RU, UK

Published in the United States of America by Cambridge University Press, New York

www.cambridge.org
Information on this title: www.cambridge.org/9780521153478

First published 2006
First paperback printing 2010

A catalogue record for this publication is available from the British Library

ISBN 978-0-521-85878-6 Hardback
ISBN 978-0-521-15347-8 Paperback

Contents

v

Preface

Although the phenomenon of luminescence was recognized as early as the seventeenth century, systematic observation and discussion of cathodoluminescence (commonly referred to as CL) and its application to geological problems did not take place until the middle 1960s. Interest in geological applications of cathodoluminescence expanded rapidly following introduction of the concept, resulting in publication of nearly a dozen English-language CL books. Although some of these books focus on commercial applications of CL (e.g., in the semi-conductor industry), many of them deal with geological applications. These books discuss the theoretical principles of cathodoluminescence and describe practical uses of CL to solution of a variety of geological problems; however, none focuses exclusively on applications in the field of sedimentology. Hundreds of research papers that discuss theoretical and applied aspects of cathodoluminescence were also published during this period. A significant number of these research contributions have focused on the practical uses of CL as a tool for studying sedimentary rocks, particularly with regard to analysis of provenance of siliciclastic sedimentary rocks and diagenesis of both siliciclastic and carbonate sedimentary rocks.

We have attempted in this book to bring together in one volume the principal applications of cathodoluminescence imaging to study and interpretation of sedimentary rocks. The book draws heavily on information available in the published literature, as well as on our own research into cathodoluminescence applications in sedimentology. The book is divided into two parts. Part I includes an introductory chapter followed by discussion of the theoretical basis for cathodoluminescence (Chapter 2) and description of the instruments and techniques used in cathodoluminescence imaging and related analytical procedures such as trace-element analysis (Chapter 3). Part II focuses on applications of CL in the field of sedimentology, the principal concern of this book. Chapter 4 discusses the use of CL as a tool for interpreting the provenance of siliciclastic sedimentary rocks. Chapter 5 evaluates the effectiveness of CL imaging for identifying and interpreting diagenetic minerals and fabrics in siliciclastic sedimentary rocks. Chapter 6 explores the CL characteristics of carbonate minerals and the usefulness of CL for description and interpretation of the diagenetic features of carbonate sedimentary rocks. Finally, Chapter 7 explores applications of CL imaging to a variety of miscellaneous topics: skeletal petrology, apatite, sedimentary ore deposits, petroleum geology, archeology, and Precambrian rocks.

Cathodoluminescence microscopes and scanning electron microscopes and electron probe microanalyzers equipped with CL detectors are commonly available instruments in many university and commercial laboratories. Therefore, many graduate and undergraduate students, as well as academic and industry professionals, have access to CL facilities. We hope that this book will be of use to students and professionals alike who may be interested in exploring the many exciting applications of cathodoluminescence imaging to sedimentological problems.

Acknowledgments

Examples and illustrations used in this book have been drawn from a variety of published sources as well as from our own work. We wish to thank Abbas Seyedolali and Young-In Kwon for assistance with the scanning electron microscope in acquisition of cathodoluminescence images and for stimulating discussions of cathodoluminescence applications. Gordon Goles (deceased) shared his insight into the theoretical aspects of cathodoluminescence; his incisive grasp of pertinent principles has been an inspiration. Kari Bassett read some early chapters and provided critical feedback. Finally, we thank Patricia Corcoran for reviewing the entire manuscript and providing expert editorial assistance.

1
Introduction

Many minerals emit radiation, referred to as **luminescence**, when bombarded by an energy source. Emissions are commonly in the visible range; however, ultraviolet (UV) and infrared (IR) emissions may also occur (Marshall, 1988, p. 1). Luminescence is given different names depending upon the energy source: e.g., bombardment by high-energy UV photons generates photoluminescence; a beam of energetic ions produces ionoluminescence; X-rays generate radioluminescence; and bombardment by high-energy electrons produces cathodoluminescence (Pagel *et al.*, 2000a). **Cathodoluminescence** refers to emission of characteristic visible (and UV) luminescence by a substance that is under bombardment by electrons, where the cathode is the source of the electrons. Note: the word cathodoluminescence is often abbreviated to CL.

The phenomenon of luminescence was recognized as early as the seventeenth century (Leverenz, 1968); however, systematic observations and discussion of cathodoluminescence did not take place until around 1965 (e.g., Smith and Stenstrom, 1965). Early cathodoluminescence studies were carried out with a cathodoluminescence microscope, which is fundamentally a petrographic microscope to which some kind of cathode gun is attached. Subsequently, the electron-probe microanalyzer and, especially, the scanning electron microscope have been utilized to generate high-resolution, high-magnification cathodoluminescence images (Chapter 3).

Early applications of cathodoluminescence to geological materials included observations of the CL characteristics of both carbonate and silicate minerals, particularly quartz and feldspars. Many investigators noted, for example, that some carbonate minerals display zoning in CL images, which was not visible in other kinds of images. A particularly noteworthy observation was made by Sippel (1968), who pointed out that

1

CL provides a means of distinguishing between secondary quartz, which is nonluminescent (or poorly luminescent), and primary or detrital quartz, which displays luminescence.

Interest in cathodoluminescence and its applications to geological materials has escalated sharply since these early studies. Several books have been published that deal in some way with cathodoluminescence (listed below), and hundreds of research papers have also been published. We now recognize that CL imaging can be applied effectively to a number of geological problems, including sedimentological problems. Cathodoluminescence techniques have proven to be particularly useful in provenance analysis of sandstones and shales and in study of diagenesis in both siliciclastic and carbonate sedimentary rocks. Published observations on the CL characteristics of sedimentary materials are widely scattered in the literature. In this book, we attempt to bring together material from diverse published sources, as well as from our own research, to provide a comprehensive treatment of the application of cathodoluminescence imaging to a range of geological problems involving sedimentary rocks.

Books dealing with cathodoluminescence microscopy and spectroscopy

Barker, C. E. and O. C. Kopp (eds.), 1991. *Luminescence Microscopy and Spectroscopy: Qualitative and Quantitative Applications*, SEPM Short Course, 25 (published by the Society for Sedimentary Geology, Tulsa).

A multi-author volume that covers both fundamentals (limited) and applications of cathodoluminescence. Topic coverage includes applications to carbonate rocks, sandstones, oil shales and coals, diagenetic minerals, petroleum geology, and ore deposits.

Gorobets, B. S. and A. A. Rogozhin, 2002. *Luminescent Spectra of Minerals: Reference Book,* Moscow, RPC VIMS.

Translated from the Russian by B. S. Gorobets and A. Girnis. Not widely available.

Götze, J., 2000. Quartz and silica as guide to provenance in sediments and sedimentary rocks. In *Contributions to Sedimentary Geology*, **21** (published by Schweizerbart'sche Verlagsbuchhandlung, Stuttgart).

This short monograph provides a comprehensive treatment of the characteristics of quartz, including cathodoluminescence, as provenance indicators. Includes color plates of CL images.

Götze, J., 2000. Cathodoluminescence microscopy and spectroscopy in applied mineralogy. Ph.D. Thesis, Technische Universität Bergakademie Freiberg.
This thesis covers the basic principles of cathodoluminescence; CL characteristics of quartz, feldspars, and zircons; and various applications, particularly applications to industrial materials.

Marfunin, A. S., 1979. *Spectroscopy, Luminescence and Radiation Centers in Minerals.* Berlin, Springer-Verlag.
This book covers various kinds of spectroscopy, including luminescence spectroscopy and luminescence centers. No direct application to sedimentary geology.

Marshall, D. J., 1988. *Cathodoluminescence of Geologic Materials,* Boston, Unwin Hyman.
This volume is an excellent book that provides a useful introduction to cathodoluminescence techniques and applications to geological materials. It has fairly limited coverage of applications to sedimentary rocks.

Ozawa, L, 1990. *Cathodoluminescence: Theory and Applications,* Tokyo, Kodansha.
The focus is on commercial applications of the principles of cathodoluminescence. The book contains no discussion of geological materials.

Pagel, M., V. Barbin, P. Blanc, and D. Ohnenstetter (eds.), 2000. *Cathodoluminescence in Geosciences,* Berlin, Springer-Verlag.
This work is another useful multi-author volume. It is an outgrowth of a conference on cathodoluminescence applications held in Nancy, France in 1996. All of the papers in the book are by conference participants. It is an excellent, up-to-date book; however, only a few of the papers deal specifically, and in detail, with application to sedimentary rocks.

Redmond, G., L. Balk, and D. J. Marshall (eds.), 1995. *Luminescence,* Scanning Microscopy Supplement 9, Proceedings of the 13th Pfefferkorn Conference, Scanning Microscopy International, Chicago.

This volume contains twenty-four papers, ten of which deal with mineral luminescence, eight with semiconductors, and six with experimental techniques. Cathodoluminescence is the focus of ten papers; however, only two of these discuss applications to sedimentology (carbonate rocks).

Yacobi, B. G. and D. B. Holt, 1990. *Cathodoluminescence Microscopy of Inorganic Solids,* New York, NY, Plenum Press.
This book focuses on application of cathodoluminescence techniques in the assessment of optical and electronic properties of inorganic solids, such as semiconductors, phosphors, ceramics, and minerals. Although it contains useful information about the fundamentals and techniques of cathodoluminescence, it contains no discussion of applications to geological materials.

Zinkernagel, U., 1978. *Cathodoluminescence of Quartz and its Application to Sandstone Petrology,* Stuttgart, Schweizerbart'sche Verlagsbuchhandlung.
An early book that describes cathodoluminescence imaging by using a CL attachment to a light microscope. It focuses on CL color as a tool for provenance analysis and other applications.

PART 1

Principles and instrumentation

The theoretical underpinnings of cathodoluminescence are complex and not fully understood. It is necessary, however, to develop some understanding of the causes of cathodoluminescence and the instrumentation required to generate cathodoluminescence images before moving on to discuss the application of CL to study of sedimentary rocks. Chapter 1 is a general introduction to the phenomenon of cathodoluminescence and

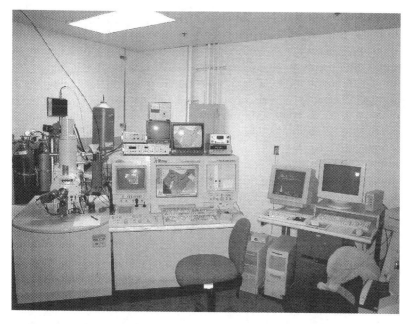

Scanning electron microscope equipped with a cathodoluminescence detector.

includes brief descriptions of some relevant books that deal with cathodo-luminescence microscopy and spectroscopy. Chapter 2 provides an introduction to the fundamental causes of CL and the factors that affect the nature and intensity of CL emissions. Chapter 3 describes the instruments and techniques currently available for CL imaging and related analyses.

2
Cathodoluminescence and its causes

Introduction

When a crystal such as quartz or feldspar is bombarded by a stream of high-energy electrons in a scanning electron microscope or other suitable instrument, photons ("particles" of light) are emitted, a phenomenon called cathodoluminescence (CL). Much is known about the origin of CL in artificial crystals because of their economic importance in the manufacture of television screens, computer monitors, and the like. Although less is known about the origin of CL in naturally occurring minerals, the fundamental causes of CL emissions are moderately well understood.

To visualize the factors responsible for CL emissions, it is useful to consider atoms in crystals in terms of the band theory of solids. Energy states of electrons in crystals depend upon whether the electrons are bound in particular atoms (inner-shell electrons) or are delocalized. Delocalized electrons are electrons that are not associated with individual atoms or identifiable chemical bonds, but are shared collectively by all the constituent atoms or ions of a substance. Delocalized electrons have wave functions (in quantum mechanics, a complex function of time and position) that in effect traverse the entire crystal. Individual atoms have discrete energy states that are associated with the orbits of shells of electrons in the atom. When atoms are spaced far apart, as in a gas, they have very little influence upon each other. By contrast, atoms within a solid, such as quartz or feldspar, have a marked effect upon other atoms in the crystal (atomic orbitals combine to form molecular orbitals). Because of the very large number of atoms that interact in a solid material, the energy levels are so closely spaced that they form bands.

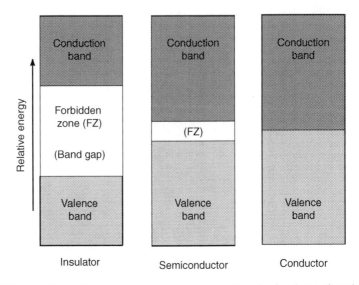

Fig. 2.1. Illustration of energy bands in minerals. Note that insulator minerals such as quartz have a wide band gap, semiconductors have a narrow band gap, and conductors have essentially no band gap.

That is, the close proximity of atoms in a solid causes the discrete energy levels of individual atoms to broaden into **energy bands**.

These energy bands are commonly designated as the **valence band** and the **conduction band** (Figure 2.1). The valence band, which has lower energy than the conduction band, is analogous to the highest *occupied* molecular orbital in a molecule and the conduction band is analogous to the lowest *unoccupied* molecular orbital in a molecule. Electrons in the conduction band are easily removed by application of an electric field. If a material has a large number of electrons in the conduction band, it acts as a good conductor of electricity. The lower-energy valence band contains a series of energy levels containing valence electrons. Electrons in this band are more tightly bound to the individual atom than are the electrons in the conduction band; however, the electrons in the valence band can be moved to the conduction band with the application of energy. In the case of cathodoluminescence, this energy is supplied by bombardment with high-energy electrons. There are more bands below the valence band, but they are not important to the understanding of CL theory and will not be discussed.

In insulators, such as quartz, a forbidden zone exists between the valence band and the conduction band, which is referred to as the **band** gap. Electrons do not reside permanently in the forbidden zone; however, they can travel back and forth through it. The band gap is quite wide in insulators. It is much narrower in semiconductors, and is absent in conductors, as shown in Figure 2.1.

Fundamental causes of cathodoluminescence

In insulator minerals, all electrons are present in the lower-energy valence band; that is, the conduction band is empty. If sufficient energy is applied to the mineral, electrons can be promoted from the valence band to the conduction band, leaving behind a so-called "hole." (Figure 2.2B). Holes can be visualized as positive charges with the effective mass of an electron. Once energized electrons are in the conduction band, they remain only a very short time before they lose energy and return to the valence band. In an ideal crystal, nothing would be present in the band gap to hinder the return of de-energizing electrons as they fall (in energy) from the conduction band to holes in the valence band. Of course, all crystals have defects of some kind, which occupy discrete energy levels in the band gap. There are a great many kinds of defects (to be discussed), which fall into the general categories of impurity ions and lattice defects. These defects constitute electron traps that momentarily intercept and hold electrons as they move through the band gap to the valence band, shown schematically in Figure 2.2 by pairs of short horizontal lines in the band gap. These traps are normally unoccupied by electrons.

Consider now what happens when a beam of energetic electrons, such as from a scanning electron microscope (SEM), impinges on a crystal such as quartz or feldspar. The energy of the primary beam is partitioned in various ways. Some of the energy is converted into X-rays, some appears as backscattered electrons of relatively high energy, some as secondary electrons of much smaller energies, and some as Auger-process electrons, also of small energies (Figure 2.3). Much of the energy is absorbed and transferred to generation of phonons, with consequent release of heat. Phonons are tiny packets of vibrational (nonradiative) energy associated with heat (e.g., Wolfe, 1998).

A little of the total energy carried in the beam acts to promote non-localized electrons from the valence band to the conduction band, as schematically indicated in Figure 2.2B, leaving holes behind in the valence band. That is, the electrons go from the ground state to an excited state. Even a small amount of the total energy applied to the

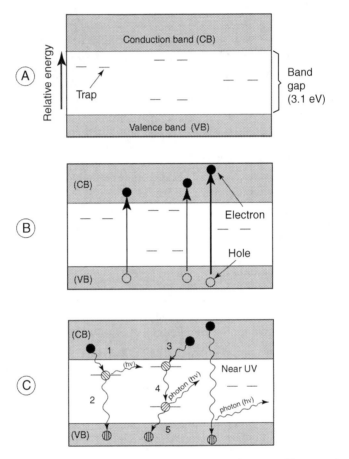

Fig. 2.2. Schematic representation of the processes that cause CL generation in minerals, as illustrated for quartz. (After Boggs *et al.*, 2001, Identification of shocked quartz by scanning cathodoluminescence imaging. *Meteoritics and Planetary Sciences*, **36**, Fig. 2, p. 785. Reproduced by permission.)

valence band may be sufficient to promote many electrons into the conduction band. After a short time, these promoted electrons undergo de-excitation and return to a lower-energy state, moving randomly through the crystal structure until they encounter a trap. (The various kinds of traps are discussed subsequently.) Electrons remain in traps only a very short time before vacating the traps, with concomitant emission of photons, and return to the ground state in the valence band (Figure 2.2C). As indicated in Figure 2.2C, electrons may encounter a single trap or multiple traps at they move through the band gap. The presence of these traps, at discrete energy levels within the band gap, is a precondition for emission of photons

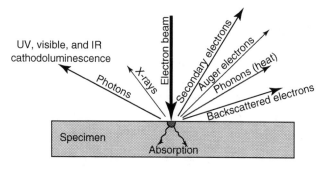

Fig. 2.3. Schematic representation of the effects produced by electron-beam interaction with a specimen in the scanning electron microscope. (After Walker, Burley, 1991. Luminescence petrography and spectroscopic studies of diagenetic minerals. In Barker, C. E. and O. C. Kopp, eds., *Luminescence Microscopy and Spectroscopy: Qualitative and Quantitative Applications*, SEPM Short Course 25, Fig. 1, p. 84. Reproduced by permission of SEPM.)

(cathodoluminescence) in the visible light range. If no traps are present, electrons fall directly back to the valence band and emit photons with wavelengths in the near ultraviolet. Residence times of electrons in traps are variable; however, most traps empty rapidly, on a timescale of microseconds. Those traps that empty promptly, producing photons with energies in the near-UV and visible portions of the electromagnetic spectrum, are the basis for cathodoluminescence. It follows from the preceding discussion, that the greater the number of electron traps present in a crystal the greater will be the number of CL emissions in the visible range.

Note that it is not necessary that photons be emitted during return of promoted electrons to lower-energy states. An alternative is emission of phonons. Phonons are packets of energy that appear as vibrations in the crystal structure, as mentioned. The energy in phonons ultimately appears as slight heating of the crystal, not as CL. Phonon emission, in competition with photon emission, is known to be facilitated by increasing densities of defects, because of overlap of the wave functions of electrons in closely packed traps. If the energy gap between the excited level and an adjacent lower-energy level is small, the electrons in the excited state tend to decay nonradiatively by phonon emission, rather than radiatively by emission of a photon. Photon emission (luminescence) occurs only if the gap to the adjacent lower level is larger than some critical value (Machel *et al.*, 1991). If a series of closely spaced traps are present, at successively lower energy levels, a promoted electron can cascade nonradiatively (phonon emission) from one energy level to another and reach ground state without emission of photons.

Kinds of luminescence centers

As mentioned, the emission of photons from a mineral undergoing bombardment by high-energy electrons is related to the presence of electron traps within the band gap between the conduction and valence energy bands. De-energizing electrons, falling back from the excited state in the conduction band to the valence band, are attracted and held momentarily by these traps. Some of the energy lost when electrons vacate traps, and continue their fall to the valence band, is converted into photons. These electron traps, which are involved in the emission of CL, are referred to as luminescence **centers**. Two fundamental kinds of luminescence centers are recognized: (1) **extrinsic centers**, often called "impurity centers," which are incorporated into the crystal owing to some property of the fluid or melt from which a mineral crystallizes, and (2) **intrinsic centers**, which result from lattice imperfections and which are often referred to as 'defect centers' (Walker and Burley, 1991). Differences in these two types of luminescence centers are examined below.

Extrinsic luminescence centers

Luminescence centers include any kind of point defect or clustered defect in a crystal structure that can absorb energy and emit optical photons (Tarashchan and Waychunas, 1995). The most abundant kinds of luminescence centers are activator centers. **Activators** are ions of various valences that substitute for cations in the host structure, e.g., Mn^{2+} can substitute for Ca^{2+} and Mg^{2+} in carbonate minerals. They have ground states that have only partially filled electron bands (orbitals) and the adjacent electron levels are separated by gaps that correspond to photons of visible or infrared light (Machel *et al.*, 1991). Tarashchan and Waychunas (1995) state that the most common activator ions are:

1 Transition metals (d–d electronic transitions), especially Cr^{3+} (oxides, silicates), Mn^{2+} (present in all types of minerals), and Fe^{3+} (silicates).
2 Rare earths with f–f electronic transitions (Pr^{3+}, Nd^{3+}, Sm^{3+}, Eu^{3+}, Gd^{3+}, Tb^{3+}, Dy^{3+}, Ho^{3+}, Er^{3+}, Yb^{3+}), and with d–f electronic transitions (Ce^{3+}, Sm^{2+}, Eu^{2+}, Yb^{2+}: both types found in fluorides, phosphates, sulfates, tungstates, silicates, and oxides).
3 Actinides, mainly U^{6+} in fluorite and carbonates.

4 Mercury and mercury-like metals, Hg^+, Tl^+, Pb^+, Pb^{2+} (silicates, sulfates).

The most common and best understood of these centers are the 3d group ions (transition metals), particularly Mn^{2+}, which is generally considered to be the most ubiquitous of all luminescence centers in minerals (Gorobets and Walker, 1995).

Some impurity ions may not be activator ions themselves but have the ability to absorb energy and subsequently transmit it to activator ions. Such ions are called **sensitizers** because they enhance the efficiency of an activator ion to generate CL. For example, the luminescence of Mn^{2+} in calcite may be sensitized by Pb^{2+} according to Tarashchan and Waychunas (1995).

On the other hand, some impurity ions that are present along with activator ions can suppress the emission of CL by activator ions because they trap part or all of the energy absorbed by activator ions. Such impurities are called **quenchers**, and the suppression process is referred to as quenching. Excited quenchers decay by multiphonon emission rather than by emission of visible light. The energy levels of these ions, particularly the gap between the ground state and the first excited state, are too close to one another to facilitate radiative transitions in the visible range (i.e., emission of CL). Therefore, phonons are emitted rather than CL (e.g., Marshall, 1988, p. 11; Machel *et al.*, 1991). Quenchers thus cause a decrease in CL emissions. If they are present in high enough concentrations, they may extinguish all luminescence.

Ions that appear to be particularly effective as quenchers include Fe^{3+}, Fe^{2+}, Co^{2+}, and Ni^{2+} (Machel *et al.*, 1991; Tarashchan and Waychunas, 1995). Nonetheless, it is apparently possible for some quencher ions (such as Fe^{3+} in feldspars) to act as luminescence centers in the infrared when present in low concentrations; Fe^{2+}, Co^{2+}, and Ni^{2+} can also give rise to luminescenece in the infrared at low concentrations, particularly at low temperatures (Gorobets and Walker, 1995). Machel *et al.* (1991) suggest that Fe^{3+} is a quencher in some minerals, including carbonates, but an activator in others. Quenching can apparently also occur by interaction of activator ions themselves at high concentrations. This quenching is due to energy transfer between nearby ions of the same type, where recipient ions are in a state in which they can decay nonradiatively. This kind of quenching is referred to as **concentration quenching** or self quenching (Tarashchan and Waychunas, 1995). Activators and quenchers are further discussed in subsequent chapters, particularly Chapter 6.

Intrinsic luminescence centers

Luminescence centers may arise also owing to the presence of a variety of defects in crystal structure. Marshall (1988, p. 8) suggests that intrinsic luminescence can be enhanced by the following factors: (1) non-stoichiometry – the state of a mineral not having exactly the correct elemental proportions; (2) structural imperfections – owing to poor ordering, radiation damage, or shock damage; and (3) impurities (non-activators), substitutional or interstitial, that distort the lattice. Götze *et al.* (2001) indicate that lattice defects can be classified according to their structure and size as follows: point defects (most important for luminescence studies), translations (crystal gliding), inclusions of paragenetic minerals, and gas/liquid inclusions.

Many point defects are so-called **paramagnetic defects**, which are characterized by unpaired electrons in their electronic configurations (J. Götze, 2004, personal communication). The presence of these defects in a crystal can be detected by electron paramagnetic resonance (EPR) study (discussed in Chapter 3). [Note: not all lattice defects are paramagnetic and detectable by ESR. In many cases (especially for trace elements), it is possible to convert non-paramagnetic (diamegnetic) defects (precursors) into paramagnetic ones by irradiation, etc.] An example of paramagnetic defects in crystals is provided by the study of quartz. More than 50 different types of paramagnetic defect centers have been reported from quartz (Weil, 1984, 1993). Paramagnetic defect centers can be divided into two main types (Götze *et al.* 2001): (1) defects caused by the presence of foreign ions (foreign-ion centers and interstitial defects), and (2) centers associated with vacant oxygen or silicon positions. Figure 2.4 illustrates some of these defects in quartz.

According to Götze *et al.* (2001), cations that readily substitute for silica include Al^{3+}, Ga^{3+}, Fe^{3+}, Ge^{4+}, Ti^{4+}, and P^{5+}. Additional cations such as H^+, Li^+, Na^+, K^+, Cu^+, and Ag^+ can be incorporated in interlattice positions. One of the most common centers in quartz is $[AlO_4]^0$, caused by substitution of Al^{3+} for Si^{4+} with an electron hole at one of the four nearest O^{2-} ions, forming O^-. The precursor state for this center is the diamagnetic $[AlO_4 / M^+]^0$ center associated with an adjacent charge-compensating cation M^+ (where $M^+ = H^+, Li^+, Na^+$). Other common paramagnetic trace-element centers in quartz are $[FeO_4 / M^+]^0$, $[GeO_4 / M^+]^0$, and $[TiO_4 / M^+]^0$.

Centers associated with vacant oxygen or silicon positions include what is referred to as E' and O^- centers. The E' center is an oxygen vacancy center; the oxygen tetrahedra are transformed into a planar

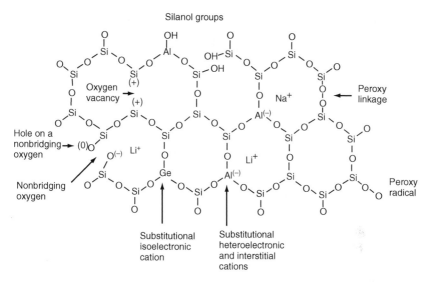

Fig. 2.4. Schematic quartz structure showing the most common intrinsic and extrinsic lattice defects. [From Götze *et al.*, 2001. Origin, spectral characteristics and practical applications of the cathodoluminescence (CL) of quartz – a review. *Mineralogy and Petrology*, **71**, Fig. 1, p. 229. Reproduced by permission.]

arrangement of three oxygen ions. The O^- centers represent different types of defect electrons on O^{2-} in tetrahedra with a silicon vacancy. The nonbridging oxygen hole center (NBOHC:$\equiv$Si–O) consists of a hole trapped in a single oxygen atom bound to a single silicon on three oxygen atoms in the SiO_2 structure (see Götze *et al.*, 2001). Oxygen excess centers include the peroxy radical ($\equiv$Si–O–O), an oxygen-associated hole center consisting of an O_2^- ion bonded to a single silicon on three oxygen atoms, and the peroxy linkage ($\equiv$Si–O–O–Si$\equiv$). OH^- centers consist of a proton bound on a regular lattice O_2^- ion, located between two O_2^- ions of the silica tetrahedron. Because of the negative net charge, additional trivalent substitutes of the Si^{4+} positions occur as charge compensation (e.g., Al^{3+}) in such a way that an additional proton is bound on the Al^{3+}.

Information about the kinds and distributions of defects in minerals such as quartz and feldspar may be obtained by various micro-characterization techniques involving CL microscopy and spectroscopy, as well as other techniques such as electron spin resonance (e.g., Stevens Kalceff and Phillips, 1995; Stevens Kalceff *et al.*, 2000). Some of the more common techniques are discussed in Chapter 3. Information about intrinsic and extrinsic emission centers has been reported in the literature by numerous researchers. Some of the emission data obtained by study of

SiO_2 polymorphs is compiled in Table 2.1; see also Stevens Kalceff et al., (2000, Table 1). Note in Table 2.1 that emissions are identified both by wavelength (in nm) and energy (in electron volts, eV) of the photon. An inverse relationship exists between energy and wavelength of photons, as illustrated in Figure 2.5. Similar data are available for some other minerals. See, for example, emission data for feldspars in Götze et al. (2000; Table 1).

The energy and corresponding wavelength of CL emissions is related to the energy level of the impurity ions responsible for the emissions (ions such as Mn^{2+} with unfilled outer shells); however, it is also influenced by the presence of other ions (the crystal field). Thus the energy and wavelength of CL emissions is not a characteristic of the ion alone; it is a property of the material (Marshall, 1988, p. 10). Other factors may also influence the wavelength of emissions. For example, Mn^{2+} substituting for Ca^{2+} in dolomite yields an emission band at 590 nm whereas Mn^{2+} substituting for Mg^{2+} in dolomite has a band at 676 nm (Machel et al., 1991).

Redmond et al. (1992) point out that CL spectroscopy studies of emission wavelength have two major objectives. By analogy to X-ray spectrometry with the electron probe microanalyzer (EPMA), the first objective is to identify trace elements in minerals. According to these authors, "the energy of CL lines or bands expresses the difference in energy between the levels located within the band gap involved in the CL emission mechanisms. The problem consists in relating the photon energy distribution, i.e., the energy levels in the energy diagram, with the nature of the luminescent centers whether intrinsic (lattice defects) or extrinsic (impurities) in origin. The problem is very complex because there is no unique relation between the color or the CL emission and the nature of the impurities within the minerals."

The second objective more or less bypasses identification of the nature of the defects and focuses instead on using the CL signal as a signature of the genetic conditions of the mineral. An example of this focus is using the CL color of quartz in sandstones as an indicator of provenance. This second objective is the principal goal of this book, which focuses primarily on the use of CL miscroscopy as a tool for studying and interpreting the characteristics and origins of sedimentary rocks.

Summary statement

Cathodoluminescence refers to the emission of characteristic luminescence by a substance that is under bombardment by high-energy

Table 2.1. *Characteristic emission bands in the luminescence spectra of quartz: CL cathodoluminescence, TL thermoluminescence, PL photoluminescence, RL radioluminescence (data from Stevens Kalceff et al., 2000).*

Emission	Proposed origin	Method
175 nm (7.1 eV)	Intrinsic emission of pure SiO_2	CL
290 nm (4.3 eV)	Oxygen-deficient center (ODC)	PL, CL
340 nm (3.65 eV)	Oxygen vacancy	TL
	$[AlO_4/Li^+]$ center	CL
	$[TiO_4/Li^+]$ center	TL
380–390 nm (3.1–3.3 eV)	$[AlO_4/M^+]$ center	RL
	$M^+ = Li^+, Na^+, H^+$	CL
	$[H_3O_4]^0$ hole center	TL
420 nm (2.95 eV)	Intrinsic emission	CL
450 nm (2.8 eV)	Intrinsic defect	CL, RL
	Self-trapped exciton (STE)	CL
500 nm (2.5 eV)	$[AlO_4]^0$ center	CL
	Extrinsic emission	RL
	$[AlO_4]^0$, $[GeO_4/M^+]^0$ center	TL
	$[AlO_4/M^+]$ center	TL
	$M^+ = Li^+, Na^+, H^+$	CL
580 nm (2.1 eV)	Oxygen vacancy	TL
	Self-trapped exciton (STE)	CL
	E' center	CL
620–650 nm (1.95–1.91 eV)	Nonbridging oxygen hole center, NBOHC	PL
	Oxygen vacancy	CL
	NBOHC with several precursors (e.g., hydroxyl group, peroxy linkage)	CL
705 nm (1.75 eV)	Substitutional Fe^{3+}	CL

Source: Götze *et al.*, 2001, Table 1.

electrons. According to solid-state band theory, electron bombardment causes electrons within a crystal to be promoted from the lower-energy valence band to the higher-energy conduction band. When these promoted

Electromagnetic spectrum		Approximate wavelength (nm)	Energy in electron volts (eV)
Ultraviolet	Far UV	<200	>6.2
	Shortwave UV	200–300	6.2–4.1
	Midwave UV	300–350	4.1–3.5
	Longwave UV	350–400	3.5–3.1
Visible light	Violet	400–425	3.1–2.9
	Blue	425–490	2.8–2.5
	Green	490–575	2.5–2.2
	Yellow	575–585	2.2–2.1
	Orange	585–650	2.1–1.9
	Red	650–700	1.9–1.8
Infrared	Near IR	700–2500	1.8–0.5
	Far IR	>2500	<0.5

Increasing wavelength → Increasing energy →

Note: energy in electron volts (eV) = 1239.8/(wavelength in nm)
(Marshall, 1988, p. 4)

Fig. 2.5. Relationship between energy of emitted photons (in eV) and wavelength (in nm).

electrons lose energy and attempt to return to the ground state, they may be trapped momentarily (for microseconds) by extrinsic (impurity) or intrinsic (structural) defects within the band gap in wide band-gap materials (insulators) such as quartz and feldspar. Energy lost when electrons vacate traps and continue their transit to the valence band is converted into photons of light, which have characteristic wavelengths. Most photons have wavelengths in the visible light range of the electromagnetic spectrum; however, emissions may also occur in the near ultraviolet and infrared. The intensity of CL emissions is a function of the density of electron traps in a crystal – the greater the number of traps the greater the intensity. Although CL emission can provide information about both the trace-element content of a crystal and factors related to its genesis, the usefulness of CL emissions as an indicator of genetic conditions is the principal focus of this book.

3

Instrumentation and techniques

Introduction

Cathodoluminescence is generated when a suitable specimen, such as a polished thin section, is bombarded with a beam of high-energy electrons. Some of the earliest observations of cathodoluminescence were made in the early 1960s by using an electron probe microanalyzer (EPMA), commonly referred to as an electron probe or electron microprobe (e.g., Long, 1963; Long and Agrell, 1965; Stenstrom and Smith, 1964; Smith and Stenstrom, 1965). In spite of this early application to CL study, the EPMA has not subsequently been used extensively for this purpose. Although CL images can be viewed in an EPMA, the EPMA is used primarily for compositional analysis.

As reported by Marshall (1988, p. 19), designs were also being developed in the early 1960s for a simpler electron excitation source that could be attached to the stage of a standard petrographic microscope to allow cathodoluminescence observations with an optical microscope. Long and Agrell (1965) discuss the design of a cathodoluminescence microscope attachment (CMA) that utilizes a heated filament gun, and Sippel (1965) designed a CMA that utilized a cold cathode gun. Subsequently, other designs for CMAs have been produced both for commercial use and for use by various research laboratories. For example, Herzog et al. (1970) described a commercial design for a CMA called the Luminoscope, an instrument still in use today.

A major breakthrough in cathodoluminescence instrumentation came with the development of CL detectors that could be attached to a scanning electron microscope (SEM). Because of its high magnification capability and the capability of combining CL and SEM observations (e.g., backscatter and secondary imaging), SEM–CL is rapidly becoming the technique of choice for high-resolution CL imaging.

19

As discussed in Chapter 2, generation of CL is related to luminescence centers in crystals. These luminescence centers arise from the presence of either impurity ions or defects in the crystal structure. To better understand cathodoluminescence requires information about the nature of luminescence centers, that is, the kinds of structural defects and/or impurity ions present in a given crystal, the energy of the photons emitted from these centers, etc. To obtain this kind of information requires techniques and instruments in addition to those commonly used for CL observations. A variety of microcharacterization techniques are in use for trace-element (impurity-ion) analysis including: EPMA, secondary ion mass spectrometry (SIMS), laser-ablation–inductively coupled plasma mass spectrometry (LA–ICP-MS), and particle-induced X-ray emission (PIXE and micro-PIXE). Techniques for evaluating defects in crystal structures are also available (e.g., electron paramagnetic resonance, EPR). This chapter focuses on discussion of standard techniques of cathodoluminescence imaging currently in use. Other pertinent microcharacterization techniques are discussed briefly.

Cathodoluminescence imaging

Optical cathodoluminescence microscopy

Optical cathodoluminescence microscopes are optical microscopes to which a cathodoluminescence stage is attached that allows a specimen (e.g., a thin section) on the stage of the optical microscope to be bombarded by high-energy electrons from a cathode gun. The stage includes a small vacuum chamber with windows, an X–Y stage movement, and an electron gun. Most of the stages utilize a so-called cold-cathode electron gun in which electrons are directed by steady discharge from a cathode at negative high voltage to an anode at ground potential (e.g., Marshall, 1993); however, some stages are equipped with a hot-cathode gun in which a heated filament supplies electrons, which are directed at lower voltages toward the anode. The CL image generated can be viewed through the objective lens of the microscope, commonly, a standard petrographic microscope.

At least two commercially produced CL stages have been available for some time. The **Luminoscope** was originally produced by the Nuclide Corporation, Acton, Massachusetts and is now produced by the Luminoscope Corporation, Moorstown, New Jersey (Luminoscope@ Yahoo.com). The **Technosyn** stage was originally produced by

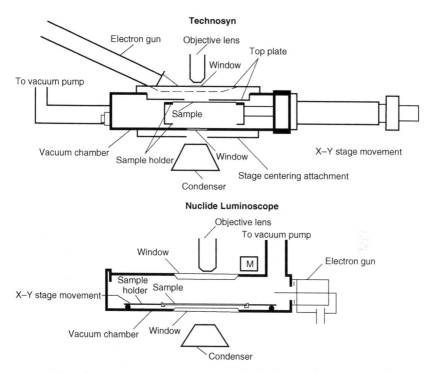

Fig. 3.1. Schematic cross-section views of the Technosyn and Luminoscope cathodoluminescence stages. (After Barker and Wood, 1986. A review of the Technosyn and Nuclide cathodoluminescence stages and their application to sedimentary geology. In Hagni R. D. (ed.), *Process Mineralogy VI*, Warrendale, PA, The Metallurgical Society, Inc., Fig. 1, p.139. Material copyrighted 1987 by TMS, The Metallurgical Society of AIME.)

Technosyn Ltd., Cambridge, UK. Successors to this stage are apparently now marketed by Cambridge Image Technology Ltd., Cambridge, UK (info@citl.com). Characteristics of early versions of these two stages have been reviewed in detail by Barker and Wood (1986) and Marshall (1993). According to Barker and Wood, the designs of the two stages include the same basic components: a vacuum chamber with upper and lower windows enclosing an X–Y stage movement with an attached cold-cathode electron source. On the other hand, the stages use different methods to direct the electron beam onto the sample (Figure 3.1). The Technosyn stage uses an electron gun aimed at the sample that is centered in the microscope optical axis. The Luminoscope stage uses deflection magnets for beam control. Both stages will accept samples of similar size.

An important difference between cold-cathode stages and hot-cathode stages, such as the SEM, exists in the vacuum within the specimen chambers. In a hot-cathode electron gun, the vacuum ranges between 10^{-5} and 10^{-6} Torr. In cold-cathode electron guns the vacuum is $\approx 10^{-2}$ Torr in order to ionize the residual gas (Redmond *et al.*, 1992). Some electrical charges in the discharge may reach the specimen and neutralize the static charge created at the surface by modifying the secondary electron yield of the specimen. Thus, conductive coatings of minerals are not necessary in cold-cathode instruments, as they are in hot-cathode instruments, including the SEM. Note: schematic details of the currently available Luminoscope stage and the Cambridge Instrument Technology Ltd (CITL) CL8200 MK5 stage may differ from those shown in Figure 3.1.

Both of these stages can be attached to a standard petrographic microscope, allowing the CL image to be viewed in color through the objective lens of the microscope. Compared to the scanning electron microscope (to be discussed), cathodoluminescence attachments to optical microscopes (CMAs) have advantages and disadvantages. Major advantages are the ability to switch quickly between CL observations and conventional optical microscope observations and the ability to see real-time, true-color display of CL patterns. Also, CMAs are relatively simple and easy to use and samples do not require a conductive coating, as mentioned. On the other hand, the range of magnifications possible with CMAs is much smaller than that of SEMs (limited by the magnification capabilities of the optical microscope). Also, the spot size of the electron beam that can be focused on a sample by using a cathodoluminescence microscope is larger than that possible with an SEM, thus limiting the resolution of CL images obtained by using CMAs. Finally, cathodoluminescence microscopes can display images composed only of photons whose wavelengths fall within the range of the visible spectrum, whereas CL images acquired in the SEM may also include contributions from photons with wavelengths in the ultraviolet and infrared portions of the spectrum.

In addition to commercially available CMAs, a number of non-commercial units are in use in various research facilities. For example, Ramseyer *et al.* (1989) describe a CMA that uses a hot-cathode electron gun. According to these authors, the high sensitivity of this instrument allows it to detect weak and short-lived luminescence in minerals with virtually no deterioration of luminescence during observation. By contrast, the high beam current density in cold-cathode CMAs leads, within a short time, to changes in luminescence intensity and color.

Scanning cathodoluminescence microscopy

Principles of SEM–CL imaging

The scanning electron microscope is an instrument capable of producing high-resolution images electronically rather than optically. A beam of electrons, focused by a system of electromagnetic lenses is used to produce enlarged images of small objects, which can be displayed on a fluorescent screen. A very fine "probe" of electrons with energies ranging from a few hundred electron volts (eV) to tens of keV is focused at the surface of a specimen. A scanning raster deflects the electron beam so that it scans the surface of the specimen in a pattern of parallel lines. The scan is synchronized with that of the cathode-ray tube, allowing a picture to be built up of the scanned area of the sample.

The concept of the SEM goes back to the 1930s. The development history of the SEM during the period from the 1930s through the 1960s is traced by Oatley (1972; 1982) and Breton (1999). The first CL images obtained in a SEM were apparently made by McMullan and Smith (Smith, 1956), as reported by Muir and Grant (1974). Muir and Grant also reported that the technique of observing cathodoluminescence in the SEM was described by Thornton (1968) and the first experimental results were published by Williams and Yoffe (1968), in a study of stacking faults in ZnSe single crystals. The very first application of SEM–CL to geological materials is difficult to ascertain; however, Thornton (1968, p. 301) mentioned that geological possibilities were being explored, but provided no details. Redmond *et al.* (1970) described the use of the SEM in cathodoluminescence observations on natural samples, and Krinsley and Hyde (1971) reported the use of SEM–CL to study fractures in quartz (sand) grains. Today, SEM–CL is widely used today for a variety of geological applications.

Figure 3.2 illustrates schematically the principal features of a modern SEM. The major components are the electron column and the control console (see photograph on page 5). The electron column consists of an electron gun and two or more electromagnetic lenses, which influence the paths of electrons traveling down an evacuated tube (e.g., Goldstein *et al.*, 2003, p. 21). A vacuum pump at the base of the column can create a vacuum of approximately one-billionth atmospheric pressure. The control console consists of a cathode-ray tube (CRT) viewing screen and the knobs and computer keyboard that control the electron beam.

When a specimen on the stage of the microscope is bombarded or irradiated by the electron beam, various kinds of emissions take place (see Figure 2.3), including emission of secondary electrons, backscattered

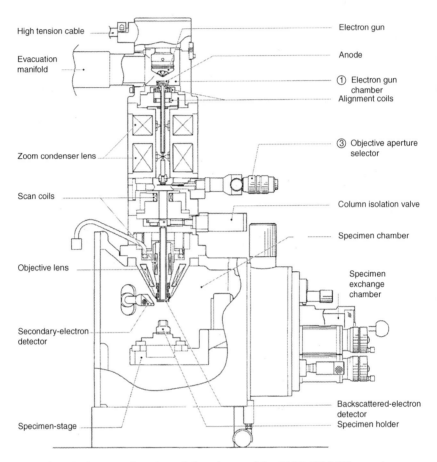

High tension cable

Evacuation manifold

Zoom condenser lens

Scan coils

Objective lens

Secondary-electron detector

Specimen-stage

Electron gun

Anode

① Electron gun chamber
Alignment coils

③ Objective aperture selector

Column isolation valve

Specimen chamber

Specimen exchange chamber

Backscattered-electron detector
Specimen holder

Fig. 3.2. Cross section view of the column of the JEOL JSM-633 scanning electron microscope. Note placement of secondary-electron and back-scattered detectors. (JEOL JSM-6300 Instruction Manual, Fig. 3.3, p. SM 63–1. Reproduced by permission.)

electrons, X-rays, and photons. Detectors must be present to capture or collect these emissions and convert the signals, by appropriate electronics, to point-by-point intensity changes on the viewing screen and produce an image (Goldstein *et al.*, 2003, p. 24). Figure 3.3 shows where secondary-electron and backscattered-electron detectors are placed to collect these emissions. The signals are collected by the detector and passed by means of a light pipe (e.g., a fiber-optic cable) to a photomultiplier tube (PMT), as illustrated in Figure 3.3. (A photomultiplier tube is a device for measuring photon counts. When light enters the photocathode, the photocathode emits photoelectrons into the vacuum.

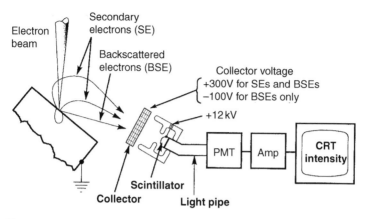

Fig. 3.3. Schematic diagram showing the positions of backscattered- and secondary-electron detectors. (After Goldstein *et al.*, 2003, *Scanning Electron Microscopy and X-Ray Microanalysis*, 3rd edn., New York, NY, Kluwer Academics/Plenum Publishers, Fig. 2.4, p. 24. Reproduced by permission.)

These photoelectrons are then directed by the focusing electrode voltages towards the electron multiplier where electrons are multiplied by the process of secondary emission. The multiplied electrons are collected by the anode as an output signal.) The signal from the PMT is amplified for display and viewing on a CRT. A detector is also present to collect X-ray signals, and a CL detector such as that shown in Figure 3.4 can be added to collect photon (cathodoluminescence) emissions.

Samples and sample preparation
A variety of different geologic sample materials can be used in SEM–CL microscopy, including rock slabs, rock chips, loose grains, and unconsolidated sediments (e.g., Trewin, 1988; Goldstein *et al.*, 2003, ch. 11). The most common specimens are polished thin sections of consolidated rock slabs. Commonly, thin sections are ground to a thickness of about 30 μm and polished to a high degree with fine diamond paste. For some purposes, sections as thick as 250 μm or more may be prepared. Some epoxy resins used to cement rock slabs to a glass slide (or to impregnate slabs that may not be completely consolidated) are luminescent. Luminescent epoxy resin can create problems in interpretation. We have found that it is generally preferable to use a nonluminescent epoxy resin such as Epotex 301-2FL, available from Epoxy Technology, or Petropoxy 154, available from Burnham Petrographics. Polished thin sections can be

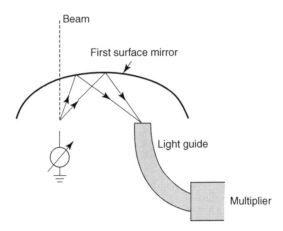

Fig. 3.4. Principal elements of a CL detector utilizing an ellipsoidal mirror. The specimen is placed at one focus and the light is reflected to the other focus, where it enters a light pipe for transmission to the photomultiplier tube. (After Goldstein *et al.*, 1992. *Scanning Electron Microscopy and X-Ray Microanalysis*, 2nd edn., New York, NY, Plenum Press, Fig. 4.28, p. 189. Reproduced by permission.)

ground from soils or unconsolidated sediments by first impregnating the sample with epoxy resin.

For some purposes, rock chips, such as cuttings from well bores, and loose grains can be glued to a slide or plug and examined in the SEM without being ground to standard thin-section thickness. An alternative technique is to grind a flat surface on the mounted grains, glue this flat surface to another slide, then grind away the first glass slide to produce a "thin section" of the grain mount. See Trewin (1988) for details of these preparation methods.

Many common minerals are nonconductive; therefore, it is necessary to coat specimens with a thin conductive coating (1–10 nm thick) to prevent charging under electron bombardment (excess beam current must flow from specimen to ground to avoid accumulation of charge). Carbon is commonly used for coating geological specimens; however, other coatings (e.g., gold) are used under some circumstances. Two methods are used to apply the coatings, which are carried out in a vacuum chamber filled with a gas such as argon or nitrogen. In the **evaporation method**, a conductive metal is heated to its vaporization temperature in a high vacuum and the evaporated metal atoms land on the specimen surface. In the **sputtering method**, a negatively charged conductive metal target is bombarded by positive ions in a low-vacuum gas discharge and the eroded metal atoms land on the specimen surface. Details of the techniques and equipment used in coating are available in Goldstein *et al.* (2003, ch. 15).

Characteristics of SEM–CL images

The electron beam of a modern SEM can be focused to a spot size as small as about 1 μm, allowing image resolution on the order of 1–5 nm (10–50 Å) (Goldstein *et al.*, 2003, p. 2). These small areas are commonly magnified on the screen from 10–10 000×; however, higher magnification is possible. The magnification of the image is simply the ratio of the area scanned to the area on the CRT. The accelerating voltages used in SEMs range from about 1 kV to >30 kV, beam currents from 1 pA to > 10 nA, and beam diameters from about 5 nm to 1 μm at the specimen. The specimen holder allows the specimen to be tilted and translated; commonly, a wide range of movements toward and away from the final lens is available, giving a free working distance from about 5 to 30 mm between the sample surface and the final lens (Pennock, 1995).

The CL image obtained by standard SEMs is a monochromatic (grayscale) image, as illustrated in Figure 3.5. The luminescence of different parts of a particular grain, such as the quartz grain in Figure 3.5, can range from very weak (dark areas) to very strong (bright areas). These variations reflect variations in the density of luminescence centers in minerals. The causes of these variations are complex and are related to the crystallization and cooling history of minerals, as well as later metamorphic influences. These factors are further discussed in appropriate subsequent chapters of the book.

Although standard CL images are grayscale images, color CL images can be generated with an SEM equipped with color (red, blue, green) filters. Grayscale images captured through these filters can be combined in a computer with appropriate software, such as Adobe Photoshop, to yield color images. Such images are often referred to as "false-color" images; however, they have reasonably representative CL colors (see Boggs *et al.*, 2002). At least one commercial company (Gatan) markets a CL detector (ChromaCL) that putatively offers live color CL imaging in the scanning electron microscope (http://www.gatan.com/sem/cromacl.html). According to Gatan, ChromaCL operates by using the principle of efficient light collection from the specimen, dispersion onto an array detector, and live color mixing of the channels, which are measured simultaneously. We have not had an opportunity to test this instrument.

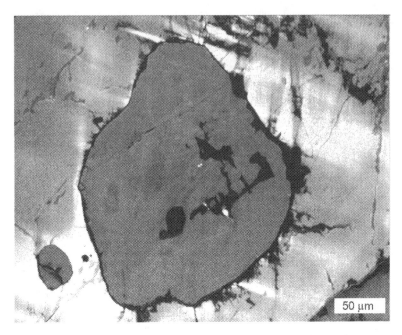

Fig. 3.5. An SEM–CL image of a plutonic quartz grain, Mt. Ashland Pluton (Jurassic), southern Oregon.

Problems with SEM–CL imaging of carbonate minerals

Scanning electron microscope CL imaging can generate high resolution, high magnification images of silicate and many other minerals; however, standard SEM–CL imaging techniques do not work well for many carbonate minerals (e.g., calcite and dolomite). Luminescence in calcite and dolomite takes a relatively long time to decay (following cessation of excitation by the electron beam), a phenomenon referred to as phosphorescence. Phosphorescence produces a highly deleterious smearing on the image as the electron beam is rastered over the area of interest (Lee *et al.*, 2005).

Special techniques are required to avoid phosphorescence when acquiring SEM–CL images of carbonate minerals. Lee (2000) showed that usable panchromatic SEM–CL images (images formed using all of the light that is emitted from a sample and recorded by the photomultiplier tube) of carbonate minerals can be formed by setting the *dwell-time* of the electron beam at each point in the raster to a value sufficiently large that luminescence from a given point contributes very little to the net signal from subsequent points. This technique prevents smearing of the image; however, considerable time (tens of minutes) may be required to acquire a high-resolution image. Reed and Milliken (2003) proposed an alternate technique, called *limited-wavelength imaging*, which requires less acquisition time. This technique takes advantage of the fact that most carbonate minerals commonly have two discrete CL emission peaks, one at long wavelengths (orange to red) and one at short wavelengths (UV to blue). Most of the phosphorescent smearing occurs with the longer-wavelength CL emissions. Reed and Milliken excised the long-wavelength (orange–red) emissions by employing an optical filter secured in front of the CL detector. They thus acquired sharp SEM–CL images by using only the UV–blue wavelength emissions. Although limited-wavelength images can be acquired using shorter dwell times than those suggested by Lee (2000), images of features (such as CL zoning) that are formed from UV–blue emissions only may not be a completely adequate proxy for panchromatic images.

Supplementary techniques in CL analysis

Backscatter imaging in the SEM

In addition to acquiring CL images with an SEM, it is generally desirable also to acquire a backscattered-electron image for comparison with the CL image. When a specimen is irradiated with high-energy electrons, backscattered electrons are emitted along with photons and X-rays (Figure 2.3). The amount of electron backscattering is indicated by the backscattering coefficient (η), which is defined as the fraction of incident electrons that do not remain in the specimen (Krinsley *et al.*, 1998): $\eta =$ number of backscattered electrons/number of incident electrons. The electron-backscattering coefficient increases with increasing atomic number (Z) of the elements. The number of backscattered electrons emitted, and thus the intensity of backscattered emission from a mineral, is a function of the average or mean atomic number ($\bar{Z}$) of the mineral. It follows from this that the grayscale intensity of a backscattered image, as it appears on a CRT, is a function of the average atomic number of the mineral. That is, low intensity of backscattered emissions yields dark images and high intensity of backscattered emissions yields bright images. For example, quartz, with a moderately low $\bar{Z}$, appears dark in back-scattered images and pyrite, with a high $\bar{Z}$, appears very bright. Thus, backscattered images are extremely useful for differentiating different mineral phases when grains are being selected for CL examination or to compare different mineral phases within a single grain undergoing examination by CL imaging. Application of backscattered-electron microscopy to study of sedimentary rocks is discussed in detail by Krinsley *et al.* (1998).

Energy-dispersive X-ray spectroscopy (EDS)

When an electron beam such as that from an SEM interacts with a sample, electrons are "promoted" from a lower-energy shell to a higher-energy shell (a process referred to as excitation), leaving a so-called "hole" in the lower-energy shell. When an electron then drops from a higher-energy shell back to a lower-energy shell to fill this hole, X-rays are emitted. Inasmuch as the energy of X-ray photons is related to the energy difference between electron shells, the X-ray photons are characteristic of the elements present in the specimen undergoing analysis. X-rays emitted by each chemical element thus have a unique energy; consequently, they are referred to as characteristic X-rays. By collecting characteristic X-rays emitted from a sample, we can obtain compositional information in terms of the atomic species present (e.g., Russ, 1984). This process is referred to as energy-dispersive X-ray spectroscopy (EDS).

EDS with a cathodoluminescence microscope Characteristic X-rays can be collected with a suitable detector attached to a cathodoluminescence microscope and displayed as an EDS spectrum, that is, a plot of X-ray energy versus intensity. One of the earliest EDS detectors designed for a CL microscope was that of Marshall *et al.* (1988); see also Marshall (1991). This detector, when attached to a Luminoscope (Figure 3.6), allowed collection of EDS spectra such as that shown in Figure 3.7. Because each element generates X-rays with a characteristic energy, each peak in an EDS spectrum identifies a particular element. Furthermore, the relative intensities of the peaks reflect the relative abundance of each element in a mineral. Therefore, EDS spectra provide largely qualitative elemental compositional data.

A problem with this early EDS detector was the large spot size of the electron beam, about $1.5 \times 1.0 \, \text{mm}$ (Marshall, 1991). Thus, it was difficult to get an accurate elemental analysis of very small grains. This problem has been mitigated to some extent in newer detectors, such as that marketed by Cambridge Image Technology, Ltd., which have spot sizes less than $100 \, \mu\text{m}$ (e.g., Vortisch *et al.*, 2003).

EDS with a scanning electron microscope Energy-disperse X-ray spectroscopy is an extremely important adjunct to SEM–CL analysis. It

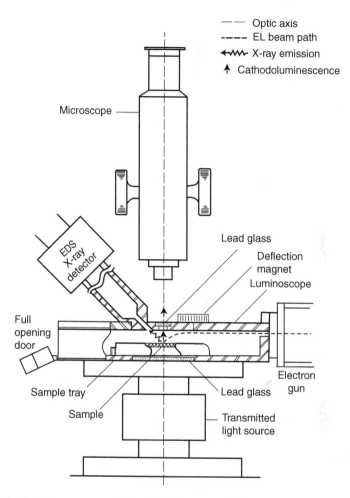

Fig. 3.6. Arrangement of the EDS detector on the Luminoscope. (After Marshall, 1991. Combined cathodoluminescence and energy dispersive spectroscopy. In Barker, C.E. and O.C. Kopp, eds., *Luminescence Microscopy: Quantitative and Qualitative Aspects*, SEPM Short Course 25, Fig. 3, p. 29. Reproduced by permission of SEPM.)

allows chemical analysis (commonly qualitative), and thus relatively unambiguous identification of mineral grains, so that an operator can be confident of the identity of grains selected for SEM–CL imaging. Furthermore, owing to the high magnification and resolution available with an SEM, it is also possible to obtain chemical information about small areas within individual grains that may be of interest because of CL observations or backscattered imaging.

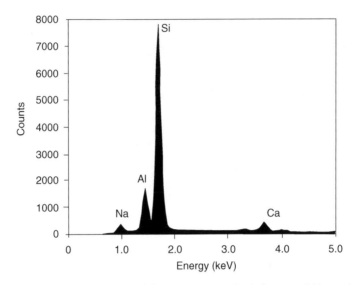

Fig. 3.7. Example of an EDS spectrum acquired from a feldspar in Westerly, Rhode Island, USA granite. After Marshall, 1991. Combined cathodoluminescence and energy dispersive spectroscopy. In Barker, C. E. and O. C. Kopp, eds., *Luminescence Microscopy: Quantitative and Qualitative Aspects*, SEPM Short Course 25, Fig. 8, p. 31. Reproduced by permission of SEPM.)

The basic features of an SEM–EDS system are illustrated in Figure 3.8. X-Rays emitted from a sample undergoing irradiation by an electron beam are captured by a detector and passed through a preamplifier. The signal then goes to a main amplifier and from there to an analog–digital converter. From the converter, it is sent to a computer that displays the X-ray spectrum on a CRT (Goldstein *et al.*, 2003, p. 299).

A variety of commercial EDS systems are currently available. Operation of the EDS system is controlled by an on-screen menu such as that shown in Figure 3.9, which also illustrates how the spectra are displayed. Peak energies can be determined by a dial or mouse-controlled feature know as a cursor. The energy of any peak and the corresponding number of counts can be read directly from the numerical display. The peaks can be labeled to show the chemical element responsible for the peak. This can be done automatically for all peaks or single peaks may be labeled by clicking on them. For many purposes in CL analysis, determining that a particular element is present (qualitative analysis) is adequate. It is not necessary to know the concentration of that element in the sample. Some EDS systems do permit quantitative analysis of samples if standards are available.

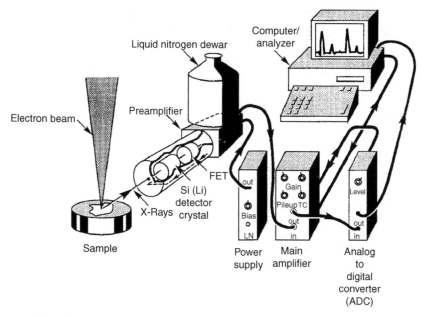

Fig. 3.8. Schematic representation of an energy-dispersive spectrometer and associated electronics. (From Goldstein, J. I. *et al.*, 2003, *Scanning Electron Microscopy and X-Ray Microanalysis*, 3rd edn., New York, NY, Kluwer Academic/Plenum Publishers, Fig. 7.2, p. 299. Reproduced by permission.)

X-Rays can also be captured and displayed by using wavelength-dispersive spectroscopy (WDS). Wavelength-dispersive spectrometers are used to select X-rays of interest for analysis. Selection is made by Rayleigh scattering of the X-rays from a crystal located between the sample and X-ray detector. By changing the angle of incidence, an analytical crystal can be made to diffract X-rays of different wavelengths. The resulting X-rays are counted by using X-ray detectors that must be moved to accommodate the changing incident angles on the crystal. A typical qualitative analysis thus involves obtaining a recording of X-ray intensity as a function of crystal angle. It is possible to convert peak positions to wavelengths through Bragg's law and then use the Moseley relationship (see Goldstein *et al.*, 2003, p. 279) to identify the elemental constituents. The spectra can be displayed as plots of intensity versus wavelength (in Angstroms) or as plots of intensity versus both wavelength and energy (in keV). Wavelength-dispersive spectroscopy commonly gives more accurate results than does EDS; however, it is generally more expensive, time-consuming, and difficult to use. Some

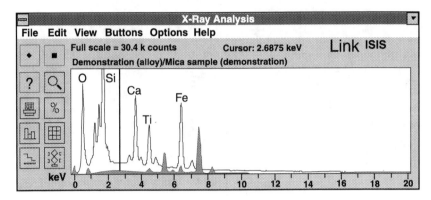

Fig. 3.9. Features of an EDS CRT display showing the computer menu that controls the system and the X-ray spectra generated (counts plotted vs. peak energies in keV). Note that this particular ERD systems allows spectra from two samples (shaded and unshaded) to be displayed together for comparison purposes. Note also that each peak can be labeled with the symbol of the corresponding chemical element. (After Link ISIS Operators Guide, Rev. 3.0, vol. 1, X-ray Analysis, Oxford Instruments, p. 70. Reproduced by permission.)

analytical systems allow both EDS and WDS to be used. Wavelength-dispersive spectroscopy is the principal mode of X-ray measurement for the electron probe microanalyzer (EPMA). Details of both EDS and WDS are provided by Goldstein *et al.* (2003).

Cathodoluminescence spectroscopy

Cathodoluminescence microscopes, such as the Luminoscope, permit direct observation of color CL images from uncoated specimens, which avoids CL color absorption by the coating. As subsequently discussed in Chapter 4, the cathodoluminescence color of quartz has been used extensively as a tool for provenance analysis, especially of sandstones, based on the assumption that quartz from different primary sources displays different CL colors. Unfortunately, visual perception of color and its application to provenance analysis is a subjective process. Visual observation can be enhanced by addition of an accessory that can collect photons and displays spectra of CL intensity versus wavelength. These instruments are referred to as CL spectrometers or CL **spectrophotometers**. Analysis of CL spectra provides a more objective basis for evaluating CL color than does visual observation. Furthermore, luminescence spectroscopy provides valuable additional information about factors such as luminescence decay times, lattice occupancy, and the nature of luminescence centers (e.g.,

Walker and Burley, 1991; Götze, 2002). Attempts have even been made to use CL spectroscopy for direct quantitative determination of selected trace elements in minerals (e.g., Haberman *et al.*, 1998).

The basic components of a CL spectrophotometer include (1) the source of the emissions (photons emitted from a sample examined in a CL microscope, SEM, or EPMA), (2) a monochromator that takes the continuum of emissions and selects a single wavelength of emission or a band of wavelengths to be analyzed by the detector, (3) the detector, e.g. a photomultiplier tube, which monitors variations in the intensity of the emissions and converts the photon signal into an electrical signal, and (4) the signal processor, which takes the electrical signal and processes it into a magnitude and format that can be displayed by a readout device. Commonly, the CL spectra are obtained by dispersing the CL signal of the area of interest by a grating spectrometer. Many spectrometers in use employ monochromators that project the light signal to a dispersion grating, transforming the polychromatic beam into a spectrum. This spectrum is moved along a narrow exit slit of the spectrometer by turning the grating, which allows detection of the intensities of separate (mono-chromatic) spectral bands by a photomultiplier (Habermann *et al.*, 2000a). Other kinds of spectrophotometers are also in use, including those with a prism monochromator (e.g., Habermann *et al.*, 2000a; Marshall, 1993; Redmond *et al.*, 1992; 2000). For example, one type of spectrograph involves use of diode arrays. The diode array (OMA: Optical Multichannel Analyzer) system disperses CL light across a flat field occupied by an array of hundreds of diodes that collect the entire spectrum at once (e.g., Owen, 1991; Redmond *et al.*, 1992; 2000).

Readout devices display the spectra in the form of relative intensity (e.g., counts) of CL emissions versus wavelength in nm (or photon energy in keV, if desired). Figure 3.10 shows a typical emission spectrum and indicates the activator ions responsible for the major emissions. Intensity and wavelength calibration of spectrophotometers is necessary for CL spectra to have substantial validity (e.g., Owen, 1991), and raw spectral data commonly require correction for the response function of the spectrophotometer (Redmond *et al.*, 1992).

Spectrophotometers have been designed for use both with cath-odoluminescence microscopes and electron microscopes, including the microprobe (EPMA) and the.scanning electron microscope. Several commercial photospectrometers are available, including some designed specifically for CL spectroscopy with the scanning electron microscope. Figure 3.11 shows Gatan Incorporation's MonoCL spectrophotometer

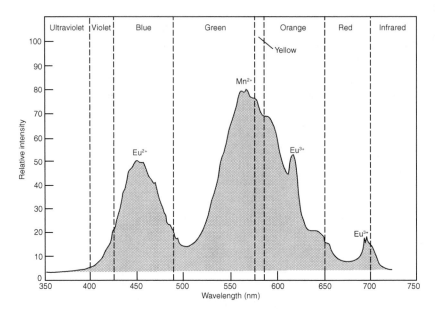

Fig. 3.10. Emission spectrum from an apatite crystal. The activator ions responsible for the emission peaks are manganese (Mn^{2+}) and europium (Eu^{2+} and Eu^{3+}). To the eye, the CL emission is blue with some yellow areas. (Based on Mariano, 1988, Fig. 8.7.)

attached to a scanning electron microscope. Cathodoluminescence emissions are affected by temperature and are commonly much more intense at lower temperatures. Therefore, some CL systems are equipped with cold stages that allow temperature of the specimen to be controlled (e.g., as in Figure 3.11).

The main purpose of this book is to explore the application of CL techniques to study of various sedimentological problems. Many of these techniques involve observation of fabrics and textures revealed by CL, and interpretation of these features in terms of provenance, crystal growth, mineral zonation, and diagenesis. Although many of these CL techniques can be successfully applied in empirical studies (e.g., Seyedolali et al., 1997a), the geological significance of petrographic CL fabrics cannot be fully interpreted unless the causes of CL emission are understood (Walker and Burley, 1991; Walker, 2000). Cathodoluminescence spectroscopy with an SEM is an extremely important micro-characterization tool that allows identification and characterization of extrinsic and intrinsic defect centres. Numerous luminescence centers in silicate and carbonate minerals have been identified on the basis of photon wavelength (and energy) as revealed by emission spectroscopy

Fig. 3.11. Gatan's MonoCL3 spectrophotometer installed on an SEM, including a liquid helium cold stage. Gatan product advertisement, downloaded from the Internet on May 11, 2004. Reproduced by permission.

(e.g., Götze *et al.*, 2000; Habermann *et al.*, 2000a; Stevens Kalceff *et al.*, 2000). Several associated techniques for microcharacterization of defect centers, which complement CL microscopy, are also in common use, as discussed below.

Other microcharacterization techniques

Electron probe microanalysis (EPMA)

As mentioned, EPMAs make use of WDS to determine elemental composition of minerals. That is, X-rays emitted from a sample bombarded with a focused beam of electrons can be used to obtain a localized chemical analysis. Electron probe microanalyzers are widely available, have a small sampling volume, and data reduction procedures and quantitative calibration methods are well established. Furthermore CL can be observed through the optical microscope in the EPMA (Reed and Romanenko, 1995). Thus, small areas of interest within a CL image can be selected for analysis. All of these factors make an electron probe microanalyzer a desirable instrument for determining the trace-element (impurity ion) concentration in a mineral of interest. Unfortunately, the usefulness of EPMAs for trace-element analysis is limited by the extreme care needed when analyzing elements at very low abundances and the fact that, for many elements, even long counting times coupled with a detailed background characterization will not achieve the required sensitivity (e.g., Müller *et al.*, 2003). Thus, trace-element concentrations are commonly measurable only at the ppm level or higher. See Reed (1995) for operational details of the EPMA.

Secondary-ion mass spectrometry (SIMS)

Secondary-ion mass spectrometry involves bombarding the surface of a sample with a beam of positive ions, which causes emission of a variety of secondary particles from the sample, particularly positive and negative secondary ions. This process is referred to as **sputtering**. The emitted ions are analyzed in a mass spectrometer, which separates them by mass and displays them as a spectrum. The spectrum consists of a series of peaks of different intensities occurring at certain mass numbers. By comparing the intensities of the peaks with a standard, quantitative analysis of trace elements in a sample can be achieved with a high degree of sensitivity (low ppm to ppb level).

The SIMS instrument consists of (1) a positive-ion source, in which primary ions are generated, transported, and focused toward the sample, and (2) the mass analyzer, in which sputtered secondary ions are extracted, mass separated, and detected (e.g., Arlinghaus, 2002). The ion source is commonly positive argon (Ar) ions, O_2^+ ions, or cesium (Cs) ions; however, ion guns that produce positive ions from liquid metal – gallium (Ga), gold (Au), indium (In) – are also in use.

Three kinds of mass analyzers are used in SIMS systems: quadrupole mass spectrometers, magnetic sector field mass spectrometers, and time-of-flight mass spectrometers (see Arlinghaus, 2002 and Hutter, 2002). The time-of-flight (TOF) mass spectrometers, which have been in use since about 1985 (Steffens *et al.*, 1985), are particularly noteworthy because of their capacity to provide simultaneous detection of all masses of the same polarity. In a TOF mass analyzer (Figure 3.12), all sputtered ions are accelerated to a given potential (2–8 keV), so that all ions have the same kinetic energy. The ions are then allowed to drift through a field-free drift path of a given length, commonly 1.5 m, before striking the detector. Light (less dense) ions travel the fixed distance through the flight tube more rapidly than identically charged heavy (dense) ions. Thus, measurement of the flight time of ions with mass-to-charge ratio (m/q), provides a means of mass analysis. A sophisticated high-frequency pulsing and counting system must be used to time the flight of the ions to within a tenth of a nanosecond.

The SIMS data are displayed as a spectrum, which consists of a series of peaks of different intensities (i.e., ion current) occurring at certain

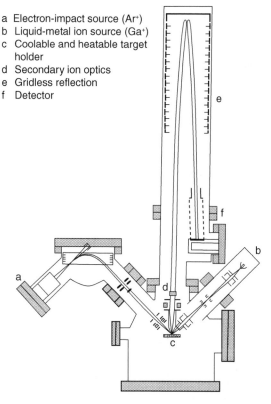

a Electron-impact source (Ar⁺)
b Liquid-metal ion source (Ga⁺)
c Coolable and heatable target holder
d Secondary ion optics
e Gridless reflection
f Detector

Fig. 3.12. Schematic diagram of the time-of-flight SIMS system used at the University of Münster, Germany. (From Arlinghaus, H. F., 2002. Ion detection. In Bubert, H. and H. Jenett, eds., *Surface and Thin Film Analysis*, Weinhein, Wiley-VCH, Fig. 3.4, p. 91. Reproduced by permission.)

mass numbers. The SIMS spectra provide evidence of all the elements present, as well as insight into molecular composition. Peak heights cannot be taken to be directly proportional to concentrations in the sample. It is necessary to use well-characterized standards to quantify trace elements.

The SIMS instrument can be operated either in the **static** mode or the **dynamic** mode. The difference in the two lies in the incident (or primary) current densities used. The current density denoting the division between static and dynamic modes is 10^{-8} to 10^{-9} A cm^{-2} (Benninghoven *et al.*, 1987); thus, current density is much higher in dynamic SIMS. Because of the low current densities used in static SIMS, only a very thin surface layer (10–50 Å) is affected by sputtering and relatively little destruction of the surface layer occurs. Dynamic SIMS is commonly used for measurement of trace elements in minerals. Because of the higher current density used, sputtering penetrates through the thin surface layer to depths of as much as 100 nm. Thus, dynamic SIMS is a more destructive technique than is static SIMS. Dynamic SIMS permits measurement of depth profiles, based on detection of the masses of interest during sputter removal of the sample material.

A major advantage of TOF SIMS is that ions of all masses can be detected. Thus, the amount of data generated in a short time is enormous. The disadvantage is that very sophisticated data acquisition systems are required to handle and process the huge amounts of data. See also Wilson *et al.* (1990) for additional discussion of the advantages and disadvantages of SIMS.

Secondary-ion mass spectrometry has been used extensively in commercial applications that involve study of thin surface films in metals, semiconductors, oxides, ceramics, biomolecules, polymers, etc. It has also been applied to materials of geologic interest for high-sensitivity analysis of samples for elements present at ppm and ppb levels of concentration (e.g., MacRae, 1995). Secondary-ion mass spectrometry and TOF-SIMS have been used to study a wide range of geologic problems, including: study of rare-earth elements in silicate minerals; analysis of zircons for U–Pb chronology; trace-element studies of marine biomineralization; determining the concentration of light elements such as Li, Be, and B in silicates; analysis of the chemical structure in coal macerals; evaluation of the boron composition of subduction-zone metamorphic rocks; identification of mineral phases on basalt surfaces; detection of radionuclides in particles from soil samples; analysis of oxygen isotopes and matrix effects in complex minerals and glasses; study of chemical weathering surfaces;

study of oxygen isotope composition of diagenetic quartz overgrowths; analysis of fluid inclusions in minerals; and evaluation of many other geologic problems. See Hinton (1995) for a more extended discussion of applications in the earth sciences.

Published studies of SIMS aimed specifically at determining trace-element (activator and quencher ions) composition in silicate or carbonate minerals as an adjunct to CL studies are relatively few. A notable exception is the research of Müller *et al.* (2003), who analyzed trace elements in quartz by combined EPMA, SIMS, laser-ablation ICP-MS, and cathodoluminescence. They report that the highest-precision data were obtained by SIMS; however, this advantage was compromised to some extent by the lack of a high-quality quartz reference sample for calibrating the technique. They also point out that SIMS has lower spatial resolution than does EPMA.

A further problem that we encountered in microcharacterization of individual quartz grains with the TOF-SIMS is difficulty in locating an analysis point precisely within the grain in relation to CL features. For example, to analyze small areas within a grain, such as the CL-dark areas in Figure 3.5, requires very precise positioning. Secondary-ion mass spectrometry instruments are capable of acquiring optical (CCD) images, ion-induced secondary-electron images, and backscattered images; however, it is difficult to compare these images with CL images. Precise measurements from a reference point, or points, within a sample are required for positioning, a difficult task.

Laser-ablation–inductively coupled plasma mass spectrometry (LA–ICP-MS)

Laser-ablation ICP-MS is an analytical technique that uses a focused laser beam to remove material from a sample for analysis, a process called ablation. A small volume of material of interest is ablated from the sample surface (probably in the form of condensed droplets) and transported in an argon carrier gas directly to the inductively coupled plasma (a state of matter that contains electrons and ionized atoms). The ions are then separated according to mass (by using a mass spectrometer), detected, multiplied, and counted by using digital electronics (e.g. Ridley and Lichte, 1998). Sensitivities (limits of detection) range from a few ppm to a few ppb.

The basic instrumental setup for LA–ICP-MS is illustrated in Figure 3.13. A sample, either a polished thin section or a rough sample, is placed

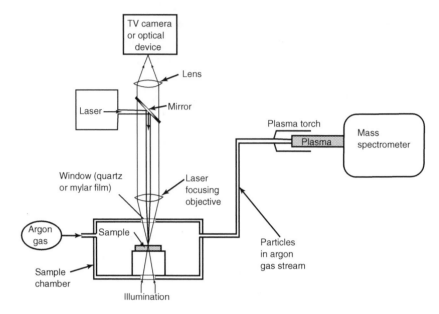

Fig. 3.13. Schematic representation of the laser-ablation–inductively coupled plasma mass spectrometer system. The system includes four principal parts: a source of laser energy, a sample chamber for laser ablation, a gas plasma torch, and a spectrometer detection system. (Based on Ridley and Lichte, 1998, Fig. 1, p. 200.)

in the sample cell, which is continuously purged with nearly pure argon gas. Samples are positioned under the laser beam and viewed with the aid of a high-resolution color video camera and monitor, or a suitable optical system. The top part of the sample cell is transparent to laser light, allowing the laser beam to be focused onto the sample surface. Particles ablated by the laser beam are carried to the argon or argon/nitrogen plasma (ICP) where a plasma torch heats the plasma to a temperature of about 10 000 K (Perkins and Pearce, 1995). There, they undergo melting, dissociation, and ionization prior to detection of specific masses within the mass spectrometer. The spectrometer used in LA–ICP-MS is commonly a quadrupole mass spectrometer or a magnetic sector field mass spectrometer (e.g., Arlinghaus, 2002; Hutter, 2002); however, TOF mass spectrometers can also be used. The operational details of LA–ICP-MS are covered in several available publications (e.g., Jackson *et al.*, 1992; Jarvis *et al.*, 1992; Perkins and Pearce, 1995; Ridley and Lichte, 1998; Sylvester, 2001). The multiple-author volume edited by Sylvester (2001) describes numerous applications of LA–ICP-MS to problems in the earth sciences; however, few of the applications apply directly to sedimentology.

Samples for LA–ICP-MS analysis do not require any special preparation, and several different kinds of samples can be used: uncovered or polished thin sections, sawn blocks, polished blocks, or mounted mineral grains. The laser beam ablates craters ranging in diameter from about 20–50 μm. Detection limits, which are inversely proportional to the third root of the ablation volume (Perkins and Pearce, 1995), range to tenths of a ppm for many elements of geologic interest. Quantitative analysis requires the use of well-characterized internal standards (e.g., Flem *et al.*, 2002). The low detection limits combined with relatively straight forward calibration procedures (Müller *et al.*, 2003) are strong points of the LA–ICP-MS method; however, because of the large crater size created by ablation, it has poorer spatial resolution than does the EPMA and SIMS.

As is the case with SIMS, LA–ICP-MS has been applied to a variety of geologic problems where quantitative detection of trace elements is required (e.g., Perkins and Pearce, 1995; Ridley and Lichte, 1998; Sylvester, 2001). The method is also valuable in CL studies for identifying luminescence centers arising from impurity ions; however, its relatively poor spatial resolution limits its usefulness for characterizing microscale regions within individual grains, such as those shown in Figure 3.5.

Proton-induced X-ray emission (Micro-PIXE)

The scanning nuclear microprobe (proton microprobe), like the EPMA, uses a focused beam of charged particles to excite samples. Instead of electrons, however, the focused beam commonly consists of protons (Fraser, 1995). The incident proton beam is used to generate characteristic X-rays, hence the name proton-induced X-ray emission (PIXE). In early studies, an unfocused beam was used to excite the sample. Subsequent developments led to use of a focused beam as small as 1 μm or less (i.e., Micro-PIXE). The proton microprobe has higher spatial resolution and lower X-ray background than the EPMA. Thus, it is possible to use Micro-PIXE to measure trace-element concentrations down to levels of around 1 ppm on a 1 μm beam spot (Fraser, 1995).

The samples used for geological studies are commonly thick (100 μm) polished thin sections. Calibration of analyses is made by comparison with standards of similar composition, which have been analyzed by other techniques. Calibration may also be done by means of sophisticated peak-stripping and modeling software, which may allow standardless analysis (e.g., Ryan *et al.*, 1988). See also Campbell and Czamanske (1998). Operational details of the PIXE method for trace-element analysis are available in the publications cited above and in a number of other publications (e.g., Johnson *et al.*, 1995).

Micro-PIXE analyses have been applied to numerous geological problems (e.g., Campbell and Czamanske, 1998), particularly in the areas of mineralogy and petrology. In the field of sedimentology, the technique has been used to study cathodoluminescence zoning in a variety of minerals, including carbonate minerals, quartz, feldspar, apatite, titanite, and zircon. Its lower detection limits makes it a desirable alternative to EPMA analysis.

Electron paramagnetic resonance (EPR)

Electron paramagnetic resonance (EPR), also referred to as electron spin resonance (ESR) and electron magnetic resonance (EMR), is a spectrographic technique that detects chemical species that have unpaired electrons (paramagnetic centers). The process involves resonant absorption of microwave radiation by paramagnetic ions or molecules with at least one unpaired electron spin and in the presence of a static magnetic field. When a strong magnetic field is applied to a paramagnetic species, the individual magnetic moment (the torque exerted on a magnet or dipole when it is placed in a magnetic field) arising via the electron spin of the unpaired electron can be oriented either parallel or anti-parallel to the applied field. This process creates distinct energy levels for the unpaired electrons, making it possible for net absorption of electromagnetic radiation (in the form of microwaves) to occur. The condition referred to as **resonance** takes place when the magnetic field and the microwave frequency are adjusted such that the energy of the microwaves corresponds exactly to the energy difference of the pair of involved states. Some researchers have referred to this procedure as being analogous to tuning a radio dial exactly to a desired station.

An unpaired electron can be aligned by an external magnetic field such that its spin precesses about the field. The direction of the spin precession can be clockwise or counterclockwise, resulting in two energy states for the electron. Irradiation with an electromagnetic beam (microwaves) can induce transitions between these different energy states, resulting in an EPR signal. That is, EPR measures the absorption of microwave radiation by an unpaired electron when it is placed in a strong magnetic field.

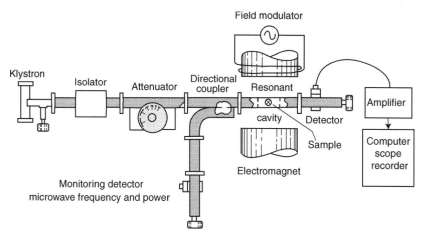

Fig. 3.14. Schematic diagram of a continuous-wave EPR spectrometer. (After Weil *et al.*, 1994. Electron Paramagnetic Resonance: *Elementary Theory and Practical Applications*, New York, NY, John Wiley & Sons, Fig. 1.3, p. 5. Copyright John Wiley & Sons, Inc. Reprinted with permission of John Wiley & Sons.)

An example of a continuous-wave EPR spectrometer, used to detect the EPR signal, is illustrated in Figure 3.14; pulse spectrometers are also in use, e.g., Schweiger and Jeschke (2001). Microwaves are generated by a klystron source, which is a vacuum tube with low-noise characteristics (Weil *et al.*, 1994, p. 5). The microwaves pass through an isolator (analogous to the monochromator in an optical spectrometer) and eventually to the resonant cavity (sample chamber), where approximately monochromatic radiation falls on the sample. Changes in intensity of the transmitted (or reflected) radiation are picked up by a suitable detector, amplified, and transmitted to a readout device (computer scope recorder).

EPR has been applied to a wide range of problems in many fields of science based upon the resonant absorption of microwaves by paramagnetic atoms, molecules, ions, and free radicals in dielectric solids, liquids, and gases. It allows study of paramagnetic impurities as trace elements and of defects formed by natural or artificial irradiation, in crystalline as well as amorphous materials. Results include locating and measuring the concentration of impurities in the compound structure, determining local crystal electric and magnetic fields and consequently local distortion around each center, suggesting charge-compensation processes, revealing atom motional effects, and identifying different types of radiation defects (Weil *et al.*, 1995).

Our concern in this book is in application of EPR to identification of defect structures in minerals that may be luminescence centers. For example, as mentioned in Chapter 2, more than 50 different types of paramagnetic defect centers have been reported from quartz alone (Weil, 1993). See also Götze *et al.* (2001). We will return to this topic, as appropriate, in subsequent chapters of the book.

PART II

Applications

Cathodoluminescence imaging has been applied to study of a variety of problems within the broad field of geology, including mineral identification, recognition of different generations of minerals, zonation of crystals, textural relationships among minerals, and brittle deformation of mineral grains. In the field of sedimentology, the principal applications have been to provenance analysis of sandstones and shales and diagenesis of both siliciclastic and carbonate sediment. Other applications include study of the outlines and internal structures of fossils not visible by light microscopy, use of CL to detect trace-element distributions in minerals, and study of apatite in sedimentary rocks. Cathodoluminescence imaging has also been used to evaluate the characteristics of sedimentary ore deposits, to study aspects of petroleum geology, and to trace the sources of archeological materials. The remainder of the book focuses on these various applications.

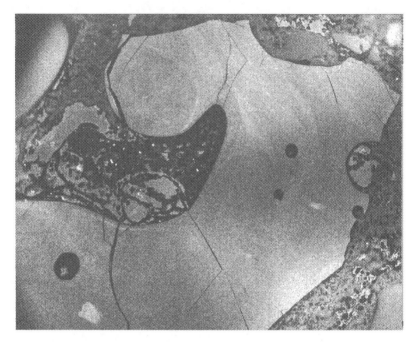

Cathodoluminescence image of a large, zoned and embayed, volcanic quartz grain, Eugene Formation (Eocene), western Oregon.

4

Provenance interpretation

Introduction

Sedimentologists have, for a great many decades, displayed a persistent interest in unraveling the provenance of siliciclastic sedimentary rocks. This interest has extended to interpretation of both the lithology of parent source rocks and the tectonic setting of source areas. Many properties of siliciclastic sedimentary rocks such as texture, sedimentary structures, chemical composition, and facies relationships may play some role in provenance analysis (e.g., Boggs, 1992, ch. 8); however, provenance analysis has focused on identification and interpretation of the particulate components (minerals and rock fragments) of conglomerates, sandstones, and, to a lesser extent, shales.

Large clasts in conglomerates can be easily and reliably identified; therefore, interpretation of source-rock lithology from study of conglomerates is relatively straightforward. That is, the coarse clasts in conglomerates can commonly be readily traced to the specific kinds of plutonic igneous, volcanic, metamorphic, or sedimentary source rocks from which they were derived. On the other hand, conglomerates probably make up less than about one percent of all sedimentary rocks; so they may not be readily available for provenance analysis in many cases. The provenance of shales, which make up roughly 50 percent of all sedimentary rocks in the geological record, is unfortunately much more difficult to assess reliably because the fine grain size of shales makes them difficult to analyze petrographically. Sandstones, which are sufficiently coarse grained to be readily studied by petrographic methods and abundant enough (about 25 percent of all sedimentary rocks) to be important, have consequently received the bulk of attention from geologists interested in provenance analysis.

Many of the minerals present in sandstones can be derived from more than one kind of source rock; therefore, a variety of techniques have been used in attempts to link minerals such as quartz, feldspars, and various heavy minerals to their sources. For example, the degree of undulatory extinction in quartz, the nature of twinning and zoning in feldspars, the trace-element composition of various heavy minerals, and the CL characteristics of quartz and feldspars have all been used as diagnostic criteria for provenance analysis. These various techniques have been reviewed in several publications (e.g., Zuffa, 1985; Boggs, 1992; Götze and Zimmerle, 2000). In this chapter, we single out cathodoluminescence and focus specifically on applications of cathodoluminescence techniques to interpretation of provenance.

Cathodoluminescence color of quartz as a provenance tool

In a pioneering paper published in 1978, Zinkernagel appears to have been the first worker to investigate the relationship between quartz cathodoluminescence characteristics (color) and the kinds of source rocks in which quartz occurs. Using a cathodoluminescence microscope to observe CL colors, Zinkernagel reported that blue to violet luminescence is characteristic of quartz in volcanic, plutonic, and some contact metamorphic rocks; brown luminescence characterizes metamorphosed igneous rocks, metasediments, some contact metamorphic rocks, and regionally metmorphosed rocks; and nonluminescing quartz indicates an authigenic origin (in sediments). Following Zinkernagel's publication, numerous other geologists investigated the relationship between cathodoluminescence color and provenance (e.g., Matter and Ramseyer, 1985; Owen and Carrozi, 1986; Ramseyer et al., 1988; Owen, 1991; Walderhaug and Rykkje, 2000; Götze et al., 2001). Although these various workers differ slightly in their interpretation of CL color versus provenance, general agreement appears to be solidifying with respect to the following relationships (Götze and Zimmerle, 2000):

blue to violet CL: plutonic quartz, as well as quartz phenocrysts in
  volcanic rock, and high-grade metamorphic quartz
red CL: matrix quartz in volcanic rocks
brown CL: quartz from regionally metamorphosed rocks
no CL or weakly luminescent: authigenic quartz
short-lived green or blue CL: hydrothermal and pegmatitic quartz

A common problem in CL studies of quartz arises owing to changes in CL color with increasing radiation of a sample, commonly a shift from blue toward red. Clearly, this shift could affect provenance interpretation. Götte *et al.* (2001) used digital image analysis to quantify the color shift in quartz grains and determine the direction and amount of shift. Applying this method allows a greater number of quartz types (up to 16) to be identified than indicated above (Richter *et al.*, 2003). Allegedly, the nature of the color shift has provenance.significance. For example: light blue → bluish violet = igneous quartz, blue → brown = metamorphic / hydrothermal quartz, dark green → brown = hydrothermal quartz.

Many geologists agree, however, that observing CL color in a cathodoluminescence microscope is a subjective process. Consequently, there has been a growing trend toward study of CL color by gathering CL spectra with a spectrometer (e.g., Marshall, 1988; Walker and Burley, 1991; Pagel et al., 2000b; Götze *et al.*, 2001). Furthermore, a significant body of work has emerged in which the investigators have attempted to link specific colors (CL emission bands) to a particular intrinsic or extrinsic defect (luminescence center), as discussed in Chapter 2.

As mentioned by Owen (1991), quartz commonly luminesces in two broad spectral peaks in the range of 440–480 nm (blue) and 620–660 nm (red), and the intensity of the blue peak generally exceeds that of the red. Perceived color, as viewed in a cathodoluminescence microscope, is largely a function of relative intensities of the blue and red emissions rather than major shifts in peak wavelength. Thus, the CL color of quartz is light blue if the relative intensity of the blue peak is strong and reddish-brown if the relative intensity of the blue peak is significantly reduced.

Detailed spectral analysis reveals, however, that CL emissions in quartz can occur at numerous wavelengths and energies (e.g., Stevens Kalceff *et al.*, 2000). Emission peaks at 175 nm, 380–390 nm, 420 nm, 450 nm, 500 nm, 580 nm, 620–650 nm, and 705 nm appear to be particularly important (see Table 2.1, Chapter 2). Information drawn from the work of numerous investigators (referenced in Stevens Kalceff *et al.*, 2000 and Götze *et al.*, 2001), who acquired data by using CL spectroscopy and a variety of other analytical techniques such as electron paramagnetic resonance (EPR), allows identification of many of the intrinsic and extrinsic defects responsible for each of these emission peaks (Table 2.1). Most CL colors of quartz are caused by structural (intrinsic) defects; activation by trace elements (e.g., Ti, Fe) plays a minor role (Richter *et al.*, 2003). It is by no means certain, however, that the exact causes of the luminescence centers in quartz responsible for each of these emissions are fully known.

In a broad sense, variations in CL color of quartz of different origins reflect variations in chemical composition of magmas and in magmatic or metamorphic fluids, in temperature, and probably in pressure, crystal growth rates, and strain rates. What is of special significance to our purposes in this chapter is the relationship of CL emission peaks to provenance. Does a particular emission peak, or combination of peaks, identify quartz from a particular kind of source rock? Relatively few empirical data appear to be available to answer this question; however, Table 4.1 summarizes some of what is known about the relationship between wavelength of CL peak emissions and quartz types. Note that considerable overlap is evident in this listing; few (if any?) CL emission peaks uniquely identify quartz from a particular kind of source rock.

Because of the subjectivity involved in observing CL color with a cathodoluminescence microscope, Boggs *et al.* (2002) explored a variation of CL spectrometry by using blue (380–515 nm), green (515–590 nm), and red (590–780 nm) optical filters, mounted in a CL detector attached to a scanning electron microscope, to obtain CL images of quartz of known provenance. This procedure allowed us to capture digital (grayscale) CL images at three specific wavelength ranges of the color spectrum, as well as to acquire an unfiltered image (at wavelengths ranging from ~200–700 nm). These grayscale images were then analyzed to obtain a quantitative measurement of the intensity or brightness of CL emission in the blue, green, and red color bands by using Adobe Photoshop software. The threegrayscale images can be combined in Photoshop to create a so-called false-color image, if desired. Details of the technique are given in Boggs *et al.* (2002).

We examined quartz from numerous samples of volcanic, plutonic igneous, and metamorphic rocks, as well as from hydrothermal veins. By comparing the intensity (luminosity) of blue, green, and red quartz CL emissions from these samples, we were able to make a quantitative evaluation of the CL color of quartz from all of these sources. Figure 4.1 presents the basic results. Volcanic quartz (Figure 4.1A) shows very strong emissions in the blue wavelength range and appears bright, light blue in composite ("false") color images. Plutonic quartz (Figure 4.1B) displays a range of emissions from very strong in the blue wavelength range to moderately strong in the red wavelength range. The perceived color of plutonic quartz ranges from bright blue to reddish-brown in composite color images. Metamorphic quartz (Figure 4.1C) displays a

Table 4.1. *Relationship between peak CL emissions in quartz and quartz provenance*

Emission wavelength (nm)	Kind of defect center	Provenance interpretation
175 (UV)	Intrinsic (unidentified)	?
290 (UV)	Intrinsic (unidentified)	?
340 (UV)	Intrinsic (unidentified), extrinsic	?
380–390 (UV)	Intrinsic (compensated [AlO$_4$/M$^+$] centers; short-lived)	Common in hydrothermal quartz
420 (blue)	Intrinsic (unidentified)	Volcanic and granitic quartz?
450 (blue)	Intrinsic (unidentified), extrinsic	All genetic quartz types
500 (blue)	Intrinsic (compensated [AlO$_4$/M$^+$] centers; short-lived)	Pegmatitic, igneous, hydrothermal quartz
580 (blue)	Intrinsic (unidentified)	Hydrothermal?
620–650 (red)	Intrinsic (nonbridging oxygen hole center) (NBOHC)	Present in most quartz
705 (IR)	Intrinsic?	Metamorphic quartz?

Data from Götze *et al.* (2001), Richter *et al.* (2003).

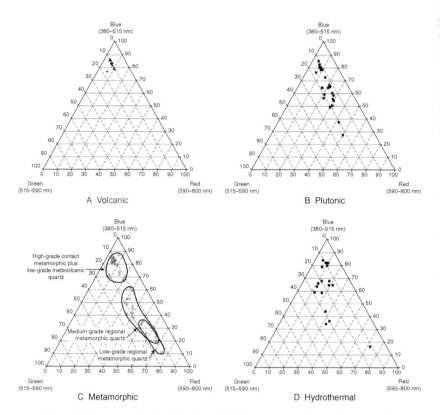

Fig. 4.1. Mean relative luminosity of CL images of volcanic (A), plutonic
(B), metamorphic (C), and hydrothermal (D) quartz acquired through red,
green, and blue filters. Relative luminosity calculated by normalizing red,
green, blue luminosity values to 100 percent. (After Boggs *et al.*, 2002. Is
quartz cathodoluminescence color a reliable provenance tool? A quantitative
examination. *Journal of Sedimentary Research*, **72**, Figures 3, 4, 5, 6, p. 413.
Reproduced by permission of SEPM (Society for Sedimentary Geology.))

wide range of emission intensities depending upon the metamorphic
grade. High-grade metamorphic quartz displays strong emissions in the
blue wavelength range, very similar to that of volcanic quartz. Medium-
grade metamorphic quartz has a larger component of red emissions, and
low-grade metamorphic quartz has an even larger component of red
emissions. Thus, the CL color of metamorphic quartz in composite color
images ranges from bright blue to reddish brown to brown. The CL
emissions from hydrothermal quartz reveal a distribution pattern of
intensity values similar to those of plutonic quartz; however, a somewhat
stronger component of green CL emissions is evident in hydrothermal

quartz. Unfortunately, the overlap in CL colors among volcanic, plutonic, metamorphic, and hydrothermal quartz casts doubt on the practicality of using CL color as a provenance indicator. Neuser *et al.* (1989) also expressed reservations about using CL color as a provenance indicator. They noted that brown or violet luminescing quartz is not necessarily of metamorphic origin, as previously believed, and concluded that "a genetic correlation of quartz particles based upon their cathodoluminescence properties is not possible, because quartz grains with similar emission spectra may have grown under very different conditions."

The unfiltered image acquired of each quartz grain allowed us to measure the total luminosity (CL intensity) of each quartz grain. By subtracting the combined numerical luminosities of the red-, green-, and blue-CL images from the total luminosity of the unfiltered image, we were able to estimate the luminosity arising from CL emission in the ultraviolet (UV) between wavelengths of ~ 200–$385\,\text{nm}$. A plot of total luminosity versus luminosity of images generated by CL emission in the near UV shows that UV luminosity is greater in plutonic quartz than in most volcanic and metamorphic quartz (Figure 4.2). Some hydrothermal quartz also displays moderately intense UV luminosity.

Intense CL emission from the UV in plutonic quartz compared to that in volcanic quartz appears to be related to temperature of quartz formation (and perhaps cooling rate?). Marshall (1988) noted that intrinsic luminescence is associated with defects; crystals formed at higher temperatures have more defects and are more brightly luminescent than are crystals of the same mineral formed more slowly at lower temperatures and with fewer defect concentrations. Thus, volcanic quartz, which commonly forms at higher temperatures and in shorter times than plutonic quartz, has higher total luminosity (brighter CL) than does plutonic quartz. Reference to Figure 2.2, Chapter 2, schematically illustrates why plutonic quartz has higher emission of CL in the UV than does more brightly luminescent (in the visible range) volcanic quartz. Promoted electrons in the conduction band (CB) that fall back directly from the CB to the valence band (VB) without encountering an electron trap, generate photons in the UV range. The fewer traps within the band gap, the less likely the electrons are to encounter a trap and generate CL in the visible-light range. Plutonic quartz, with fewer electron traps than volcanic quartz, thus tends to generate more CL in the UV range because a high percentage of promoted electrons lose energy by falling directly from the CB to the VB. Presumably, low-temperature hydrothermal quartz also generates considerable CL in the UV range for similar reasons. It may be

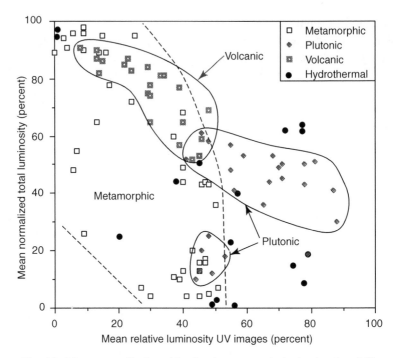

Fig. 4.2. Mean normalized total luminosity versus relative luminosity of CL images generated by emission in the UV. Relative UV luminosity is calculated by normalizing UV, red, green, blue luminosity to 100 percent. (After Boggs *et al.*, 2002. Is quartz cathodoluminescence color a reliable provenance tool? A quantitative examination. *Journal of Sedimentary Research*, **72**, Figure 9, p. 412. Reproduced by permission.)

possible to exploit the differences in ultraviolet CL luminosity between volcanic and plutonic quartz as a provenance tool; however, this is an area of study that needs further work.

Cathodoluminescence color of feldspars

Numerous studies of the CL color of feldspar have been reported in the literature, most of which focus on establishing the relationship between feldspar CL color and the intrinsic and extrinsic defect centers responsible for the color. For example, Marshall (1988, p. 112) reports that Fe^{3+} activates red CL in orthoclase; however, if other activators are also present the CL color may be blue, green, or yellow. Anorthite activated by ferrous iron (Fe^{2+}) has green luminescence, and Ti^{4+} activates blue CL in K-feldspar. Blue CL is the most frequently observed color in feldspars. Owen (1991) cites studies reporting that labradorite luminesces green and green–blue, whereas bytownite and anorthite luminesce yellow. Microcline and orthoclase are commonly pale blue, but some may be red, and sanidine is commonly blue–green. Authigenic feldspar is generally nonluminescent. Finch and Klein (1999) report that blue CL in alkali feldspars appears to relate to the presence of electron holes on bridging oxygens, particularly on the Al–O–Al bridge, as determined by EPR studies.

Perhaps the most comprehensive review of the CL colors of feldspars and the nature of the luminescence centers in feldspars is that provided by Götze *et al.* (2000). These authors list more than 25 emission bands in feldspars (Table 4.2) and identify the activators responsible for the colors, on the basis of data drawn from a host of other investigators. They conclude that the most common CL emission bands in feldspars are caused by intrinsic defects of the type $Al–O^-–Al$ (450 nm, blue, especially in K-feldspar), $O–Si \cdots M^+$ (around 500 nm, blue–green, in alkali feldspars) and the incorporation of Mn^{2+} into the M-position (*c*. 500 nm) and Fe^{3+} into the T-position (around 700 nm, red) of the feldspar structure, respectively. Other elements that can act as activators of CL emission in feldspars are Tl, Pb, Cu, some rare-earth elements, and possibly Cr. Incorporation of these activator elements into feldspars is determined by element supply and the pressure, temperature, and redox conditions during formation of the parent rock.

The CL color of feldspars provides valuable information about different feldspar phases, the distribution of alkali and plagioclase feldspars in rocks; the nature of compositional zoning; fine-scale structures, and mineral intergrowths, and identification of authigenic overgrowths (nonluminescing) on detrital feldspar grains. We can attempt to draw some generalities about provenance applications from all this; however, few studies have been reported that demonstrate a definite relationship between the CL colors of feldspars and provenance. Until sufficient empirical studies have been made, the relationship between CL color of feldspars and provenance remains tenuous.

Cathodoluminescence fabric analysis of quartz

Principles

The preceding discussion relating to application of CL color of quartz and feldspar to provenance interpretation underscores the difficulties in

Table 4.2. *Characteristic luminescence emission peaks and associated activators in feldspars. PL photoluminescence, RL radioluminescence, TL thermoluminescence, IR infrared spectroscopy, OSL optically stimulated luminescence*

Activator	Color	Peak	Method
Tl^+	UV	280 nm	PL
Pb^{2+}	UV	280 nm	RL
?	UV	330 nm	TL
			TL, IR-OSL
Ce^{3+}	UV	355 nm	CL
	Bluish green	490 nm	CL
Eu^{2+}	Blue	420 nm	CL
			TL, RL
Cu^{2+}	Blue	420 ± 5 nm	CL
			TL, RL
$Al\text{–}O^-\text{–}Al$	Blue	450–480 nm	TL, RL
			CL
Ti^{3+}	Blue	460 ± 10 nm	CL
		450 nm	CL
Ga^{3+}	Bluish green	500 nm	CL
$O^-\text{–}Si\cdots M^+$	Bluish green	500–510 nm	TL, RL
Mn^{2+}	Greenish yellow	559 nm	CL
		570 ± 5 nm	CL
		540–561 nm	CL
		550–565 nm	CL, TL
Fe^{3+}	Red/IR	700 ± 10 nm	CL
		705–730 nm	CL
		700–780 nm	CL
		690–760 nm	RL
		680–745 nm	CL
		688–740 nm	CL, TL
Sm^{3+}	Red	Several peaks	CL
	Blue–IR	Several peaks	CL
Dy^{3+}, Eu^{3+}	Blue–IR	Several peaks	CL
Tb^{3+}, Nd^{3+}		Several peaks	CL
?	IR	860 nm	CL
Cr^{3+}	IR	880 nm	RL

Data from Götze *et al.* (2000, Table 1, p. 250).

application arising from the subjectivity of judging color visually, as well as the fact that quartz (and feldspar?) derived from different kinds of source rocks can have very similar CL colors. Is there a more reliable way of utilizing the CL characteristics of minerals as a provenance tool, other than by use of CL color? It has long been known that some minerals, including quartz, display in CL images internal fabrics such as zoning and healed fractures. As mentioned in the preceding discussion of CL color, variations in the CL characteristics of quartz are an indication of conditions of origin and reflect variations in chemical composition of magmas and in magmatic or metamorphic fluids, in temperature, and probably in pressure, crystal growth rates, and strain rates. Because the conditions of origin differ for plutonic, volcanic, and metamorphic rocks, the CL fabrics of quartz of different origins also differ and may thus have significance with respect to provenance interpretation.

Krinsley and Hyde (1971) may have been the first workers to apply this concept. They used CL to examine fractures in Pleistocene glacial sands and compared fracture characteristics and patterns in these sands to those in sands from beach, dune, and weathered granite environments. Although the focus of their study was not specifically provenance analysis, their observation of the CL characteristics of fractured quartz in weathered granites certainly has provenance overtones. Subsequently, Krinsley and Tovey (1978) used CL to study dune sands from Libya and gravelly sands from England. They identified six textural characteristics on the basis of CL: (1) absence of CL; (2) lack of CL contrast; (3) narrow band structures (fractures); (4) diffuse dark areas; (5) dark areas with sharp boundaries; and (6) mottled CL texture. They suggested that these features reflect the complex geological history of the quartz grains – clearly a reference to provenance.

In a similar approach, Milliken (1994) studied the CL characteristics of quartz silt in Oligocene Frio Formation mudrocks of south Texas. She concluded on the basis of her CL studies that silt in the Frio Formation differs in origin from associated sand, most likely for reasons related to provenance and not burial diagenesis. She presented CL images of several quartz grains and identified some of the quartz as volcanic in origin.

Seyedolali *et al.* (1997a) took the concept of CL fabric analysis a step further by examining the CL fabrics of quartz from a wide variety of volcanic, plutonic, and metamorphic rocks by using SEM–CL techniques. On the basis of this study, we were able to determine that quartz can display a variety of CL fabrics, as summarized in Table 4.3. This work, and subsequent research at the University of Oregon, has demonstrated that the CL fabrics of quartz can be employed as a useful tool for provenance analysis. Details of the CL characteristics of each genetic type of quartz are discussed below.

Table 4.3. *Distinctive characteristics of quartz grains revealed by SEM–CL microscopy*

Feature	Description	Quartz source	Comments
Zoning	Typically displays oscillatory-type zoning with concentric bands oriented parallel to grain edges; less commonly, nonconcentric bands extend across grains parallel or at an angle to grain boundaries; bands range in width from 1 to >50 μm	Volcanic and plutonic rocks, hydrothermal veins	Particularly well developed in volcanic quartz; present in some plutonic quartz; zoned hydrothermal quartz commonly displays a complex pattern of small-scale zoned crystals that may include sector zoning
Healed fractures	Distinct, black (dark CL) lines (<5–>20 μm wide) strongly resembling fractures; may have diverse orientations; not visible in BSE images	Plutonic and metamorphic rocks, hydrothermal veins	Particularly common in plutonic quartz; less common in metamorphic quartz, rare in volcanic quartz
Open fractures	CL-dark or CL-bright lines (depending upon epoxy resin used in sample preparation); visible in BSE images; various thicknesses and orientations	Any kind of source rock	Late-stage features that may be present in any kind of quartz, including volcanic quartz; have little provenance significance
Deformation lamellae	CL-dark, curved–straight, semiparallel lines that occur in one or more sets with various spacings; sets may extend across entire grain	Tectonically deformed metamorphic rocks	Apparently related to line defects in crystal structure generated as a result of deformation

Feature	Description	Rock types	Notes
Shock lamellae (planar deformation features)	Thin, dark-CL lines that occur in sets within which lines are essentially parallel; two or more sets oriented at different angles may be present	Shocked crystalline and sedimentary rocks	Present in quartz from meteorite or cometary impact structures and some system boundaries; may occur in feldspars
Dark-CL streaks and patches ("spiders")	Broad (>10 μm) black (dark-CL) elongated streaks with irregular boundaries and large (~10 to >100 μm), black patches; commonly occur with healed fractures	Plutonic rocks, high-grade metamorphic rocks, hydrothermal veins	Particularly characteristic of plutonic quartz; occur rarely in high-grade metamorphic rocks; absent (or very rare) in volcanic quartz
Indistinct mottling	Irregular CL intensity across grains, producing a mottled pattern	Metamorphic rocks, some volcanic rocks	Most common in metamorphic quartz
Nondifferential (homogeneous) CL	Grains display nearly uniform CL (little difference in CL intensity)	Metamorphic rocks, some volcanic rocks	Most common in metamorphic quartz

Characteristic CL fabrics of quartz

Volcanic quartz

Volcanic quartz is particularly characterized by the presence of zoning. Some volcanic quartz displays distinct, oscillatory-type zoning, resembling in pattern the compositional zoning common in plagioclase (Figure 4.3). The zoned grains exhibit concentric light and dark bands oriented roughly parallel to grain edges. Most of the bands are thin ($<10\,\mu$m); however, the width of some bands in some crystals may exceed $50\,\mu$m. Bands are revealed by distinct to subtle differences in CL intensity (luminosity); thus boundaries between bright-CL and dark-CL bands may range from sharp to indistinct.

Not all zoned volcanic quartz displays well-developed, fine-scale concentric zoning; zoning may be coarse and ill defined and the pattern of zoning may be irregular (e.g., Figure 4.4A). Rarely, we have found volcanic quartz phenocrysts that appear to be unzoned in CL images.

Fig. 4.3. Cathodoluminescence image of an oscillatory-zoned volcanic quartz phenocryst from a volcanic clast in the Lake Hill Conglomerate (Triassic–Jurassic?), New Zealand. The black, rounded areas are melt inclusions; the thin, black lines are fractures. Photograph courtesy of Anekant Wandres and Abbas Seyedolali.

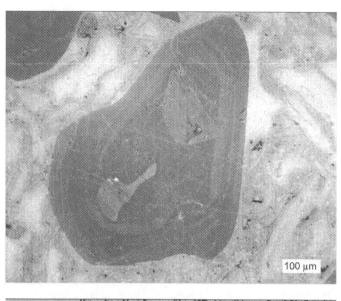

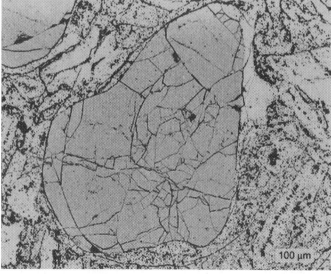

Fig. 4.4. Zoned volcanic quartz phenocryst from the Bishop Tuff (Pleisto-cene), eastern California. (A) Cathodoluminescence image. Note that the widths of the zones are greater than those shown in Figure 4.3 and the shapes of the zones are more irregular. (B) Backscattered-electron image of the same grain shown in (A). Note that the zones do not show up in the backscattered image. The thin, black lines in the backscattered image are open fractures.

Adjusting the contrast of such images with the SEM commonly shows that faint zoning is present. Most volcanic quartz that we have examined is zoned; however, Bernet and Bassett (2005) report significant amounts of unzoned volcanic quartz in some New Zealand volcanic rocks. In addition to CL zoning, volcanic quartz phenocrysts may display melt inclusions (Figure 4.3) and resorbed boundaries (note truncated zones in Figure 4.4A), which aid in their identification. Figure 4.4B is a back-scattered-electron (BSE) image of the same grain shown in Figure 4.4A. Note that the grain does not display zoning in the backscattered image, indicating that the cause of the zoning is not variations in major-element chemistry. Likewise, zoning is not visible in petrographic images. The thin dark lines in Figure 4.4B are late-stage, open fractures, which possibly formed as a result of cooling. Early-stage fractures that are subsequently healed by precipitation of silica do not show up in backscattered images. Healed fractures are uncommon in volcanic quartz.

The exact causes of CL zoning in volcanic quartz continue to be a subject of interest and debate. For example, Peppard *et al.* (2001) studied zoning in quartz crystals from the rhyolitic Bishop Tuff (Pleistocene) in California. They suggested that the bright-CL bands formed from a Ba-rich, Nb-poor rhyolitic magma. The sharpness of the bright-CL rims (which formed as a late-stage overgrowth) indicates that the magma from which the quartz phenocrysts grew changed abruptly; the bright-CL rims formed as the crystals grew while they sank through successive layers of differentiated magma; see also Anderson *et al.* (2000). Watt *et al.* (1997) carried out a particularly detailed CL study of oscillatory zoning in a porphyritic dacite from Nuit Bay, Jersey, United Kingdom, supplemented with secondary-ion mass spectrometry (SIMS) analysis to determine trace-element compositions. They concluded that oscillatory zones in quartz are formed by diffusion-dominated growth in a relatively static magma (i.e., one not undergoing rapid convective overturn and thus reserving a diffusive boundary layer at the crystal–melt interface) and are controlled by crystal growth rates and growth dynamics. They suggested that the CL emissions responsible for the zoning are related to trace-element substitution, most likely substitution of Al for Si, a coupled substitution accompanied by additional substitution of monovalent cations (H^+ and Li^+). They acknowledged, however, that the causes of CL in quartz are complex and can also be related to intrinsic defects, which makes direct correlation of high Al concentrations with strong CL emissions (bright CL) difficult.

Plutonic quartz

Figure 4.5 is a CL image of a fairly typical plutonic-quartz grain. Two characteristic CL features of plutonic quartz are revealed in this image: low-intensity (dark)-CL streaks and patches, and healed fractures. Areas of low-intensity CL in the form of broad (10–30 μm) black, irregularly shaped streaks or large (10 to >100 μm) black patches, informally dubbed "spiders," are characteristic of most plutonic quartz grains that we have examined. Most plutonic quartz also contains thin (<10 μm), distinct, black lines (CL weak or absent), which are healed fractures. A much less common feature of plutonic quartz is the presence of zoning, which is almost invariably accompanied by the presence of "spiders" and healed fractures (e.g., Figure 4.6). The presence of both spiders and healed fractures effectively distinguishes CL-zoned plutonic quartz from CL-zoned volcanic quartz, which rarely contains spiders and typically does not contain healed fractures.

Fig. 4.5. Cathodoluminescence image of a plutonic-quartz grain, which displays dark-CL streaks and patches, and healed fractures ("spiders"). Squamish Granodiorite (Cretaceous), British Columbia, Canada. Sample courtesy of J. K. Russell.

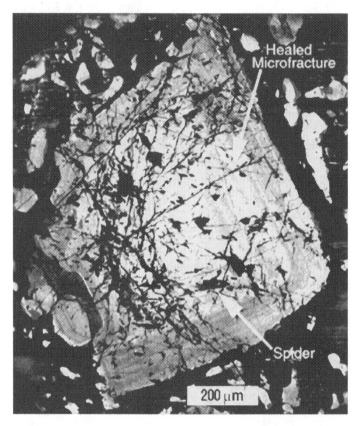

Fig. 4.6. Cathodoluminescence image of a zoned plutonic quartz from a granite clast (Jurassic–Cretaceous) dredged from the top of Yamato Rise, Japan Sea. Note abundant healed fractures and spiders. Specimen courtesy of Kensaku Tamaki.

Zoning in plutonic quartz Zoned plutonic quartz appears to be far less common than unzoned plutonic quartz, but it is common enough to be of significant interest. We have found zoned plutonic quartz in samples from such diverse sources as a granite clast dredged from the top of Yamato Rise in the Japan Sea (Figure 4.6) and small granitic stocks in the foothills of the Cascade Range, western Oregon. Among others, D'Lemos *et al.* (1997) reported zoned plutonic quartz from the Southwest Granite Complex (late Proterozoic) of Jersey, Channel Islands, UK, and Müller (2000) describes zoned plutonic quartz from the Schellerhau Granite Complex (Carboniferous), eastern Erzebirge, Germany.

Major discontinuities in CL zoning (wide bands) are probably the result of physico-chemical changes related to external factors (e.g., temperature, pressure, and magma composition). On the other hand, the development of fine-scale, oscillatory CL zoning apparently results from punctuated growth of quartz grains growing under near-equilibrium conditions; here growth is largely controlled by diffusion within the boundary layer adjacent to a growing crystal, referred to as chemical self-organization (D'Lemos *et al.*, 1997; Müller, 2000). Supersaturation of quartz in the reaction zone at the crystal-melt interface increases the quartz growth rate, leading to decrease in silica concentration if the crystal grows fast enough to exceed that rate at which SiO_2 is replenished by diffusion. Simultaneously, foreign (trace) elements accumulate in the reaction zone and boundary layer because the high growth rate favors the incorporation of impurities. The growth rate will slow when quartz growth is so fast that silica becomes depleted in the reaction zone and boundary layer (and fewer impurity elements will be incorporated). Consequently, the SiO_2 diffusion rate becomes the dominant crystal growth controlling process. The growth rate rises again as soon as the silica concentration in the reaction zone is renewed by diffusion.

The principal activator elements that account for intense CL emission in the bright bands appear to be Al and Ti; Al^{3+} is the most characteristic impurity ion that substitutes for Si^{4+} in the silicon–oxygen tetrahedra, whereas Li^+, Na^+, K^+, Fe^{2+}, and H^+ are ion compensators that enter interstitial positions. The Ti^{4+} ion is also an important substitutional ion, which creates relatively stable defect centers without interstitial charge compensators (Müller *et al.*, 2000).

A problem not addressed in the above discussion is why some plutonic quartz is zoned and some is not. Is the presence or absence of zoning related to magma composition, intrusion depth, pluton size, or other factors? Is all quartz from a particular pluton either zoned or unzoned? Finally, can the presence or absence of zoning in detrital quartz of plutonic origin be used as a guide to trace quartz to its ultimate source? There are few definite answers to these questions. One point of view holds that all plutonic quartz is initially zoned; however, quartz in large, deeply buried plutons that cool slowly eventually re-equilibrates so that the distribution of activator ions becomes homogenized and CL zoning is lost (Paul Wallace and Mark Reed, 2004, personal communication). Presumably, the quartz in smaller, shallower, faster-cooling plutons would lock in the trace-element distribution before such re-equilibration could occur. Alternatively, it is possible that the quartz in some plutons (e.g., small, hypabyssal intrusions) develops zoning whereas quartz in others (e.g., large, deep-seated intrusions) does not. Or, perhaps, quartz is unzoned at an early stage in intrusion of a deep-seated pluton but becomes zoned during ascent of magma to shallower depths. Unfortunately, the empirical data needed to support any of these hypotheses are not available, and the question of why some plutonic quartz is zoned and some is unzoned remains unanswered.

Healed fractures in plutonic quartz Healed fractures (fractures filled with secondary silica) are particularly common in plutonic quartz, are present in some metamorphic quartz, and are uncommon in volcanic quartz. The fractures appear as thin (commonly $<10\,\mu$m) black lines in CL images (e.g., Figures 4.5 and 4.6). They do not appear in backscattered images, which indicates that the fracture-healing material is SiO_2. They display very weak red luminescence in color CL images, suggesting that the fracture-healing SiO_2 precipitated slowly from fluids at temperatures well below the magmatic crystallization temperature of quartz.

As mentioned, healed fractures are particularly abundant in plutonic quartz. In CL-zoned plutonic quartz, they cut across zones, indicating that their formation postdates formation of the zones. Many fractures probably originated during the transition of beta quartz to alpha quartz, which takes place during cooling below the inversion temperature of about 570 to 600 °C. The beta–alpha transition is associated with large distortional strains, which can cause extensive cracking of the quartz grains (e.g., Sprunt and Nur, 1979; Heaney, 1994; Müller *et al.*, 2000). Further cracking can occur as a result of tectonic deformation and differential thermal and overburden stresses. Fracturing of quartz grains opens the grains to access by fluids. Subsequent precipitation of SiO_2 at lower temperatures from silica-rich fluids causes the fracture to fill and heal.

Although healed fractures are abundant in many plutonic quartz grains, they are far less common in volcanic quartz. Rapid cooling of an erupting volcanic magma may not allow time for stresses to build in volcanic quartz phenocrysts, and fracturing to occur, as the magma cools through the alpha–beta transition zone temperature. On the other hand, late-stage, *open* fractures (visible in backscattered images) are common in volcanic quartz, as illustrated in Figure 4.4B. Some late-stage breakage of quartz phenocrysts might occur owing to violent vesiculation within an erupting magma (Best and Christiansen, 1997) or open fractures may be thermally induced contraction cracks that formed as a result of rapid cooling during effusion.

CL-dark patches and streaks (spider-like fabrics) One of the most distinctive CL characteristics of plutonic quartz is the presence of irregularly shaped, CL-dark patches and streaks, which can range from < 20 to > 50 μm in length (Figure 4.5). They are almost invariably associated with healed fractures, giving them the appearance in CL images of spiders sitting on a web. They can be sparsely distributed, as in Figure 4.5 or very densely distributed, as in Figure 4.6. "Spiders" do not show up in petrographic or backscattered (SEM) images; they appear only in CL images. The fact that they cannot be seen in backscattered or petrographic images indicates that they are composed of quartz and are in optical continuity with the host quartz grain. They occur in both zoned and unzoned plutonic quartz. In zoned quartz, they commonly transect zones, as shown in Figure 4.6, indicating that they form in mature quartz crystals that are fully crystallized. That is, spiders are secondary features, like healed fractures, which form during cooling of quartz crystals, not during growth of the crystals. We have not observed these CL fabrics in undoubted volcanic quartz, although we have observed rare spiders in high-rank metamorphic quartz. Similar features have been reported in vein quartz from some porphyry copper deposits (Rusk and Reed, 2002). Thus, spiders are not confined exclusively to plutonic quartz; however, the presence of abundant spiders in detrital quartz (e.g., in sandstones), together with numerous healed fractures, most likely indicates plutonic origin.

Spider-like CL features were described by Behr as early as 1989 (Behr, 1989). We first observed them in plutonic quartz in 1997 (Seyedolali *et al.*, 1997a). Subsequently, they have been discussed by several other workers. For example, they are mentioned by Müller *et al.* (2000) in discussion of the CL characteristics of the Schellerhau Granite complex, Germany. They are described in some detail by Müller (2000), and they are discussed briefly by van den Kerkhof and Hein (2001) and van den Kerkhof *et al.*, 2001. Müller (2000) refers to them as **decrepitation traces** and reports that they are depleted in trace elements (Li, Al, K, and Ti) compared to the host quartz. Van den Kerkhof *et al.* (2001) describe them as **healing textures** and also point out that they are depleted in certain trace elements, notably Ti.

Müller (2000) suggests that the origin of decrepitation traces is related to decrepitation (breakage) of fluid inclusions induced by differences between fluid pressure and lithostatic pressure during uplift. Decrepitation generates microcracks around the fluid inclusions; presumably, fluids move outward through the microcracks, resulting in formation of defect-poor quartz (with weak emission of CL) at the cost of the defect-rich host quartz. This process is explained by the displacing of atoms along the phase boundary of the quartz with higher defect density so that the atoms fit to the lattice of the quartz with low defect density (e.g., Passchier and Trouw, 1996; Stünitz, 1998). Thus, more defect-free crystals grow at the expense of more disordered neighbors along an advancing front. This process reduces the internal free energy of the crystals involved and causes the replacement of defect centers.

An alternative explanation for the origin of "spiders" involves late-stage invasion of quartz grains by external fluids. The inversion of beta quartz to alpha quartz at temperatures of about 570–590 °C generates distortional strains that may also cause microcracking, as mentioned in the discussion of healed fractures. These microcracks could allow external fluids to gain access to the grain. The solubility of quartz at confining pressures below about 700 bar exhibits retrograde behavior between temperatures of about 590 and 350 °C (Fournier, 1999). That is, solubility increases with decreasing temperature within this temperature range. Fluids that have invaded microfractures could cause dissolution of small cavities within the quartz grains, possibly at the intersection of microfractures, as the temperature falls through the retrograde solubility zone. Additional cooling to temperatures below the retrograde zone causes solubility to decrease and results in precipitation of SiO_2 into the protospider cavities to form the spiders. This hypothesis is consistent with the fact that spiders display very weak red luminescence (appear black in grayscale images), indicating that they form at temperatures well below the magmatic crystallization temperature of quartz. Rusk and Reed (2002) propose a similar origin for spiders in vein quartz and hydrothermally altered igneous quartz. They suggest that spiders (which they refer to as **splatter and cobweb texture**) form as a result or corrosion of quartz along microfractures, followed by precipitation of CL-dark quartz in the corrosion cavities. Van den Kerkhof *et al.* (2001) also cite evidence for the formation of secondary quartz features by solution–precipitation as well as by diffusion healing.

Spiders appear to be absent, or very rare, in volcanic quartz. This scarcity may be related to the corresponding scarcity of healed fractures, which could have allowed fluid to gain access to quartz crystals. Also, fluid inclusions are less abundant in volcanic quartz than in plutonic quartz. In any case, cooling of volcanic quartz may simply take place too rapidly for the spider-forming process to proceed to completion.

Hydrothermal quartz

Although much less common than volcanic and plutonic quartz, quartz derived from hydrothermal veins can theoretically be present in sedimentary rocks as detrital grains. Therefore, we need to be aware of its CL provenance signature. Unfortunately, for the purpose of provenance analysis, hydrothermal quartz is characterized by some of the same CL features as those in volcanic and plutonic quartz, namely zoning, healed fractures, and spiders. The CL fabric of hydrothermal quartz, as it appears in quartz veins, can be extremely complex (e.g., Onasch and Vennemann, 1995; Penniston-Dorland, 2001; Rusk and Reed, 2002).

Hydrothermal quartz can display very complicated patterns of zoning, both concentric zoning and sector zoning (Figure 4.7). Sector zoning, which may be superimposed on concentric zoning, can range from simple wedge-shaped zones to very complex intrasectoral zones that can include simple tapered wedges within large sectors and fir-tree-shaped zones (Onasch and Vennemann, 1995). Vein quartz can also display healed fractures and spiders, as well as a variety of other features such as CL-bright quartz grains with CL-dark margins and jigsaw-puzzle-like

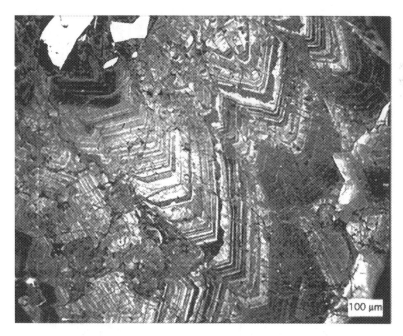

Fig. 4.7. Complex zoning in hydrothermal quartz in veins within the Butte porphyry copper deposit, Montana. Specimen courtesy of Brian Rusk.

pieces of CL-bright quartz engulfed in CL-gray quartz (e.g., Rusk and Reed, 2002). Bernet and Bassett (2005) report CL textures in hydrothermal quartz that include dark, homogeneous CL and patchy mottled CL. The CL colors of hydrothermal quartz are reported by various investigators to include blue, red, yellow, and green, and colors may be transient (last only a few seconds under the electron beam).

It is uncertain to what degree weathering and erosion of hydrothermal veins would yield sand/silt-size quartz grains in which the complex CL features of hydrothermal quartz might be recognizable and distinguishable from quartz of other origins. The abundance of hydrothermal quartz in the geological record is quite low compared to that of volcanic, plutonic, and metamorphic quartz. Therefore, hydrothermal quartz grains are unlikely to be abundant constituents of sandstones and siltstones, which is probably fortunate for the sanity of provenance analysis. We have not personably observed recognizable vein-quartz grains as detrital constituents in sandstones.

Metamorphic quartz

Metamorphic rocks can form from a variety of precursor rocks under a wide range of temperature, pressure, and interstitial fluid conditions. Therefore, it is extremely difficult to craft empirical studies of metamorphic quartz that examine the effects, on the CL characteristics of quartz, of all of the possible permutations of parent-rock composition and metamorphic grade related to temperature, pressure, and fluids. Empirical studies have shown, however, that metamorphic quartz exhibits CL characteristics that commonly allow it to be distinguished from most quartz of other origins (Seyedolali et al., 1997a).

Our observations indicate that characteristic CL signatures of volcanic and plutonic quartz are present in Precambrian rocks, which suggests that cathodoluminesence properties, once acquired, can persist in unmetamorphosed rocks for geologically long periods of time. On the other hand, quartz metamorphosed into the greenschist or higher-grade facies appears to lose its original CL properties as a result of recrystallization or annealing. We have not, for example, observed volcanic or plutonic zoning in quartz metamorphosed to temperatures of about 350 °C or higher. Bernet and Bassett (2005) report healed fractures in plutonic quartz that are partially preserved in medium-grade metamorphic rocks but disappear with increasing metamorphic grade.

Mottled-texture and homogeneous CL The two most characteristic CL features of metamorphic quartz that we have observed are mottled-texture (Figure 4.8) and homogeneous (nondifferential) CL (Figure 4.9). Mottled texture, which is very common in metamorphic quartz (and is present in some plutonic and volcanic quartz), indicates irregular distribution of activator ions or defect structures within the grains. The brighter CL areas contain more activator ions or defect structures than do the darker areas; however, little definitive work has been done to explain this distribution. Mottled texture may be caused by incomplete recrystallization or annealing accompanying metamorphism or it may be the result of deformation during metamorphism. Bernet and Bassett (2005) report a correlation between patchy/mottled CL and strong undulose extinction arising from increasing deformation. They found this CL texture in both low- to medium-grade and medium- to high-grade metamorphic quartz.

Metamorphic quartz that displays nearly uniform CL intensity (CL appears essentially homogeneous) is also moderately common. We have

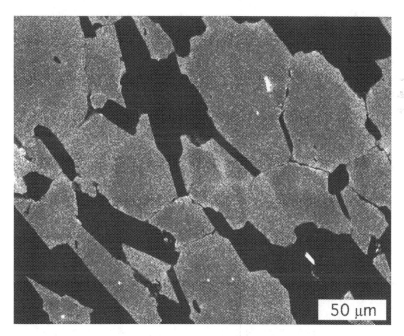

Fig. 4.8. Quartz from staurolite-kyanite-zone metamorphic rocks (Proterozoic), Snow Peak area, Idaho, showing mottled CL texture. Sample courtesy of Helen M. Land and Jack M. Rice.

Fig. 4.9. Metmorphic quartz displaying nearly homogeneous CL across the entire grains (bright grains). High-rank, contact-metamorphosed quartz from the Kangerdlugssuaq Gneiss (Precambrian), adjacent to the Skaergaard Intrusion, east Greenland. Sample courtesy of Gordon G. Goles.

observed two contrasting types of metamorphic quartz grains that display nearly homogeneous CL: (1) very bright CL grains (e.g., Figure 4.9), and (2) very dark CL grains. The bright-CL grains have high CL luminosity and appear bright blue in color images. We found this kind of homogeneous CL only in high-grade (granulite facies) metamorphic quartz, recrystallized at high temperatures. Quartz formed at very high temperatures tends to have higher luminosity than that formed under lower temperatures, consistent with the general principle that high-temperature quartz crystallizes with more structural defects and more impurity ions than does low-temperature quartz (e.g., Marshall, 1988).

We have only rarely observed metamorphic quartz grains that display homogenous but very dark CL; however, Bernet and Bassett (2005) report such grains to be very common in some metamorphic rocks of New Zealand. Intense deformation of quartz is known to reduce the intensity of luminescence. Matter and Ramseyer (1985) suggest this may occur as a result of cleaning of the quartz crystal structure from trace elements and structural defects during recrystallization. Alternatively, emission of photons may be largely replaced by emission of phonons (heat) owing to high density of defect centers created by intense deformation, thereby reducing CL luminosity (Boggs *et al.*, 2002). That is, defect structures may be so densely distributed that electrons falling back from the conduction band to the valence band (see discussion in Chapter 2) cascade from one trap to another and lose energy by emission of phonons instead of photons.

Healed fractures and deformation lamellae Healed fractures, commonly diversely oriented, are described in the preceding discussion of plutonic quartz. Such fractures are also present in some metamorphic quartz; however, they are far less common in metamorphic quartz. On the other hand, metamorphic quartz that has undergone one or more episodes of tectonic shearing may exhibit sets of CL-dark lines that superficially resemble healed fractures. For example, we have observed such features in quartz from the Bradshaw Mountains Precambrian rocks of Prescott, Arizona (Seyedolali *et al.*, 1997b). Quartz in these weakly metamorphosed metavolcanic rocks is characterized by the presence of CL-dark, curved to straight, semiparallel lines with various spacings that extend across the grains (Figure 4.10). More than one set of CL-dark lines, oriented in different directions, may be present.

These features appear to be deformation lamellae. Passchier and Trouw (1996, p. 48) report that at low-grade conditions (300–400 °C) dislocation glide and creep become important deformation processes in quartz. Deformation occurs mainly on basal glide planes and is manifested in petrographic images as sweeping undulose extinction and deformation lamellae. Apparently the lamellae are the result of *line defects* in the crystal structure, which Passchier and Trouw (1996, p. 29) describe as an "extra half lattice plane in the crystal". The quartz in some tectonically emplaced rocks may be so intensely deformed that it displays a complex, highly sheared CL fabric (Figure 4.11A). Note that the backscattered image of this grain (Figure 4.11B) does not show this sheared fabric.

Deformation lamellae are probably not healed fractures, filled with CL-dark secondary quartz. The reason why they appear in CL images as dark lines (no luminescence) is uncertain. Possibly, defect structures are so densely distributed along line defects that phonons are emitted instead of photons, as suggested in the discussion of homogeneous CL-dark grains.

Spiders in metamorphic quartz Spiders are common and abundant CL features in plutonic quartz and also form in some vein quartz, as

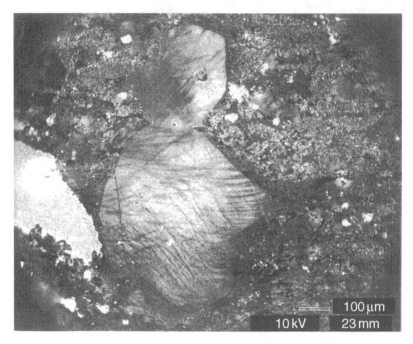

Fig. 4.10. Cathodoluminescence image of a quartz phenocryst (center) from mechanically sheared Precambrian metavolcanic rock from Prescott, Arizona. Shearing is indicated by the presence of two sets of irregular, semiparallel CL-dark lines that display various thicknesses and spacings. The different orientation of these sets suggests two different deformational events.

discussed. We have recently observed rare spiders in some high-rank metamorphic quartz, i.e., in quartz from contact-metamorphosed rocks in close proximity to the Skaergaard Intrusion, Greenland (Figure 4.12) and in granulite-facies rocks from the Adirondack Mountains, New York. Although it is possible that spiders in metamorphic quartz could be inherited from precursor plutonic quartz, relict plutonic spiders are unlikely to be preserved in high-rank metamorphic quartz. Most likely, they form in high-rank metamorphic quartz during retrograde cooling. Alpha quartz heated during metamorphism above 570–600 °C, depending upon pressure, reverts to beta quartz. Retrograde metamorphism subsequently brings temperatures back below the α–β inversion temperature, resulting in inversion, with concomitant distortional strains that could produce microfractures. Padovani et al. (1982) report extensive microcracking of amphibolite- and granulite-facies rocks during retrograde metamorphism. Although high-grade metamorphic rocks tend to be poor

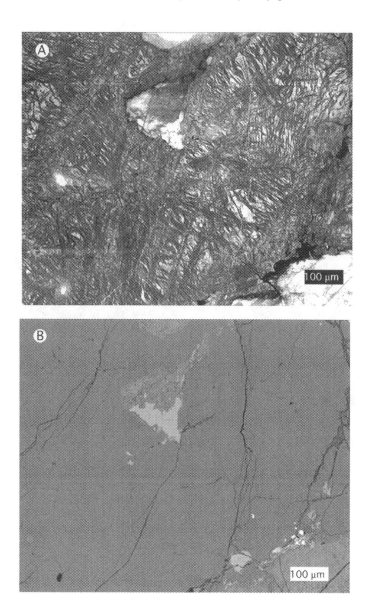

Fig. 4.11. (A) Complexly sheared quartz from tectonically emplaced trondhjemite, Wooly Creek Batholith, Klamath Mountains, northern California. Small-scale "blocky" shears are crosscut and offset by second generation of shear planes spaced 30–50 μm apart. (B) Backscattered electron image of the same grain shown in (A). Sample courtesy of M. Allan Kays.

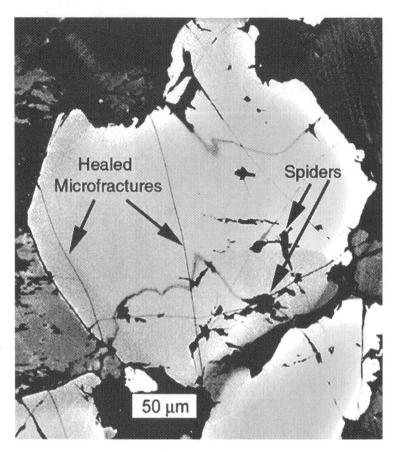

Fig. 4.12. Spiders in high-rank metamorphic quartz from contact-meta-morphosed Kangerdlugssuaq gneiss (Precambrian) adjacent to the Skaer-gaard Intrusion, east Greenland. Sample courtesy of Gordon G. Goles.

in total volatiles, fluids can be introduced from external sources, and secondary fluid inclusions may be common (Roedder, 1984, p. 362). Thus, spiders probably form in high-grade metamorphic quartz during retrograde cooling by much the same mechanisms as those discussed for generation of plutonic spiders. As far as we now know, spiders do not form in low-grade metamorphic rocks.

Cathodoluminescence textures in feldspars

Numerous studies of the CL properties of feldspars have been published; however, only a few of these studies deal with textures in feldspars revealed by cathodoluminescence imaging. We are not aware of any study that attempts to link CL textures in feldspars to provenance, analogous to CL studies of quartz. A recent paper by Götze *et al.* (2000) shows that the CL textures of feldspars can include zoning (e.g., zoning not revealed by polarizing microscopy), twinning, fractures, mineral intergrowths, alteration fabrics, such as albite replacing K-feldspar, and authigenic overgrowths. Some provenance implications can be drawn from these CL textures (e.g., zoning indicates igneous origin); however, such interpretations seem limited. Provenance-significant properties of feldspars include presence or absence of zoning and kinds of zoning, types of twinning, structural states of feldspars, and feldspar chemistry (Boggs, 1992, ch 8). These properties are commonly studied by petrographic microscopy, X-ray diffraction methods, and geochemical techniques. At this time, it appears that observation of CL textures in feldspars adds little additional provenance information beyond that available by using these conventional techniques. That perception may change with additional study of the CL textures of feldspars.

Cathodoluminescence features of zircon

Numerous studies have examined the CL characteristics of zircon (e.g., Rubatto and Gebauer, 2000; Poller, 2000; Kempe *et al.*, 2000, and references therein). The dominant CL signature of zircon is concentric (oscillatory) zoning (Figure 4.13). Zoning is apparently due to variations in abundance of uranium (U) and yttrium (Y) within the zircon crystals. Bands richer in U and Y are dark in CL images, and brighter areas are relatively poor in these trace elements (Rubatto and Gebauer, 2000). That is, CL emissions in zircon are inversely correlated to their content of U and Y. Overall, CL emission in zircon must be caused by elements other than U and probably Y; U and Y apparently act as CL quenchers, resulting in the CL-dark bands.

The main use of CL imaging of zircon has been as an adjunct to U–Pb dating of zircons. Cathodoluminescence allows identification of different types of zircon domains that then may be dated *in situ*. Dating is commonly done by using a SHRIMP (Sensitive High-Resolution Ion Microprobe) mass spectrometer, which has a spatial resolution of about 15–30 μm. By determining the ages of different zones within a zircon grain, metamorphic modifications of the grain can be differentiated from primary growth features. Determining the ages of different domains

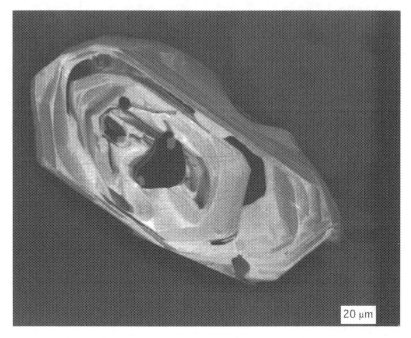

Fig. 4.13. Concentrically-zoned zircon grain from granitic rocks near Butte, Montana.

within the crystal promotes understanding of the various geological processes (magmatism, deposition, metamorphism, hydrothermal alteration, metasomatic leaching of Th, U, and Pb) recorded during the multi-stage history of the grain. Once the age(s) of a detrital zircon grain (e.g., in a sandstone) is established, the grain may be traced to a still-existing source terrane of known age.

Some CL study of zircon has aimed at establishing a link between the CL textures of zircons and their parent rocks. For example, Rubatto and Gebauer (2000) indicate that magmatic zircons commonly display oscillatory zoning (parallel to crystal faces) as well as sector zoning superimposed on the oscillatory zoning. Kempe *et al.* (2000) point out, however, that growth zoning is not restricted to magmatic environments and that primary zoning can form during some metasomatic alteration events. Metamorphic zircon have external CL zones, which are irregular in shape and which can form rims around older cores. The rims tend to show an absence of regular zoning, or show patchy, cloudy zoning. The cores of such grains may be very complex. Some workers have interpreted these rims as overgrowths; however, Kempe *et al.* (2000) state that they result from solid-state recrystallization. Thus, it may be possible to differentiate between magmatic (plutonic) zircon and zircon subsequently altered by metamorphism. Zircon can also form under hydrothermal conditions. Rubatto and Gebauer (2000) suggest that hydrothermal zircons are characterized by euhedral, elongated shape and narrow growth zones, which display relatively weak luminescence contrasts. Also, Richter *et al.* (2003) report that CL zoning and color of zircons in sandstone samples from the German Keuper Basin allows identification of a northern provenance for the Schilfsandstein (Nordic Keuper) and a southern provenance for the Burgsandstein (Vindelician Keuper).

Zircons make up less than one percent of average siliciclastic sedimentary rocks (Boggs, 1992, p. 128). To be studied effectively, they must be separated from more abundant minerals by using heavy-mineral separation techniques. The laborious tasks involved in making heavy-mineral separations and preparation of grain mounts and polished sections, discourage CL analysis of zircons as a routine provenance (source-rock) tool.

Shocked quartz

Quartz grains subjected to high-strain rate shock waves owing to meteorite or cometary impact on Earth's surface commonly display shock lamellae. Such lamellae may be present also in feldspars; however, they are much more common in quartz. Shock lamellae may occur in detrital (sedimentary) quartz as well as in igneous and metamorphic quartz. Shocked quartz is considered to be a reliable indicator of impact by extraterrestrial bodies; therefore, its presence in sedimentary rocks may be useful in the search for impact sites. Thus, in a convoluted sort of way, shocked quartz has provenance significance.

Stöffler and Langenhorst (1994) refer to shock lamellae as planar microstructures, of which two kinds are recognized: (1) planar fractures (PFs) and (2) planar deformation features (PDFs). Planar fractures form sets of parallel open fissures, with a spacing of less than $\sim$20 μm, oriented along crystallographic planes in quartz. Planar deformation features occur as multiple sets of parallel, planar optical discontinuities that take the form of thin (up to $\sim$2 μm) lamellae spaced 2–10 μm apart. Shock lamellae have commonly been studied by several methods, including petrographic microscopy, etching of samples with hydrofluoric acid followed by examination with a scanning electron microscope in the secondary mode, and by use of the transmission electron microscope (TEM).

We recently described a technique that involved application of CL imaging to study of shocked quartz from a known impact structure, Ries Crater, Germany (Boggs *et al.*, 2001). Our study showed that shock lamellae are revealed in CL images with much greater clarity and resolution than in equivalent petrographic images (Figure 4.14). As shown in Figure 4.14, shock lamellae appear as thin dark lines (exhibiting no CL) superimposed on the more brightly luminescent background of the quartz grains. Some planar features are visible in backscattered-electron (BSE) images, indicating that they are open fractures. Most do not appear in BSE images, which shows that they are not open fractures and have the same major element chemistry as the quartz grains. Shock lamellae appear dark in CL images because lamellae are filled with melt glass (Gratz *et al.*, 1996), which does not luminesce. Some lamellae may be filled with stishovite / coesite (dense quartz polymorphs formed under high-pressure conditions), which also do not luminesce (Boggs *et al.*, 2001).

Shock lamellae thus appear in CL images as thin, dark, straight, parallel lines, spaced 2–10 μm apart; two or more sets of lamellae may be present. They may superficially resemble deformation lamellae (generated by tectonism); however, deformation lamellae are more irregular in shape, orientation (may be curved), and spacing (e.g., Figure 4.10). Commonly, shock lamellae are easily distinguished from open tectonic fractures (e.g., Figure 4.15), which tend to be thicker, more diversely oriented, and more widely spaced.

Combined provenance techniques

We have focused discussion in this chapter on the application of cathodoluminescence to provenance analysis, particularly provenance analysis based on CL characteristics of quartz. Cathodoluminescence imaging of quartz is a very useful tool; however, it may not in all cases be an adequate technique for provenance interpretation when used alone. For example, CL analysis of quartz from quartz-poor rocks such as volcaniclastic sandstones may not yield a completely accurate interpretation of provenance because CL analysis of quartz does not take into account non-quartz components, which may be abundant. Many other techniques are also available for provenance evaluation (see review in Boggs, 1992, ch. 8), which can be used in conjunction with CL imaging.

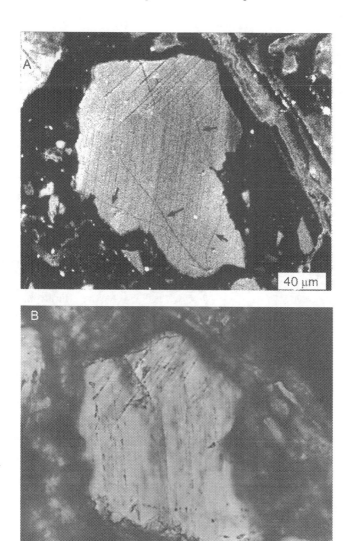

Fig. 4.14. (A) CL image of shocked quartz from Ries Crater, Germany, showing two distinct sets of shock lamellae. A few tectonic fractures (arrows) cut across the shock lamellae. (B) Optical (petrographic) image of the same grain. Note that the shock lamellae are much less distinct in the optical image. (From Boggs *et al.*, 2001. Identification of shocked quartz by scanning cathodoluminescence imaging. *Meteoritics and Planetary Sciences*, **36**, Figures 6C and 6D, p. 789. Reproduced by permission.)

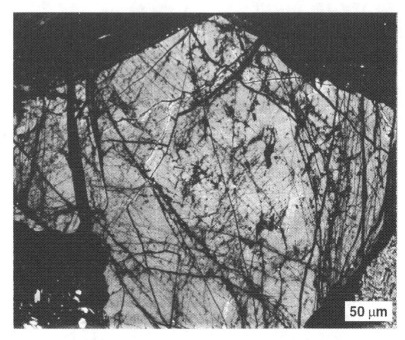

Figure 4.15. Cathodoluminescence image of a tectonically fractured quartz grain from the Bateman Formation (Eocene), Oregon Coast Range. Most of the fractures are thick, diversely oriented, and fairly widely spaced, although fractures in one set that runs diagonally across the grain (top left to bottom right) are semiparallel and more closely spaced. The thick, black fractures are filled with secondary quartz. Specimen courtesy of David Weatherby.

For example, Kwon and Boggs (2002) used both petrographic and SEM–CL techniques to evaluate the provenance of Tertiary sandstone samples obtained from drill cores from four deep wells in the Cheju Basin, northeast East China Sea, southwest of the Korean Peninsula. On the basis of petrographic data (framework mineralogy, quartz undulosity and polycrystallinity, feldspar compositions) we interpreted Cheju Basin sediments to have been derived from a mixed source terrane in which plutonic rocks were predominant, metamorphic rocks less abundant, and volcanic and sedimentary rocks least important. The SEM–CL imaging of quartz yielded data that suggested a different interpretation; SEM–CL data indicated that metamorphic quartz predominates over plutonic quartz in all samples, indicating that metamorphic rocks were the most abundant source material. On the other hand, SEM–CL revealed very little volcanic quartz even though most samples contained modest amounts of volcanic rock fragments, thereby undervaluing the importance of volcanic rocks as a sediment source. These results suggest that petrographic methods are less well suited for distinguishing between plutonic and metamorphic quartz than is SEM–CL; however, SEM–CL imaging does not adequately document volcanic contributions. The two techniques used together yield a more realistic interpretation of source-rock lithology than does either technique alone.

We did not attempt in the Cheju Basin study to examine the same grains by both petrographic and SEM–CL methods. Bernet and Bassett (2005) took the evaluation process a step further by doing exactly that. They examined the optical properties of quartz grains in quartz-rich sedimentary rocks of New Zealand and compared these properties to the features revealed by SEM–CL imaging of the same grains. This technique allowed them directly to compare petrographic features such as undulatory extinction to features of the same grains revealed in CL images. They conclude that combining techniques reduces ambiguities inherent in either technique used alone.

Bernet and Bassett's integrated technique is a bit time consuming in that a grain is first identified by using one technique (e.g., in a petrographic image), then the sample must be transferred to another instrument (the SEM) and the same grain identified and studied in that instrument. This task is accomplished somewhat more easily if the work is carried out with a cathodoluminescence microscope (CMA), where both a petrographic image and a CL image can be viewed in the same instrument. Vortisch *et al.* (2003) describe the use of this technique (but not for provenance analysis) in the study of sandstones from Sweden and Germany. The advantage of using the cathodoluminescence microscope lies in the ease of switching back and forth between petrographic and CL imaging. The disadvantage is poorer resolution and magnification of images compared to those acquired by use of the SEM.

5

Cathodoluminescence characteristics of diagenetic minerals and fabrics in siliciclastic sedimentary rocks

Introduction

Diagenesis involves both physical and chemical processes that act to modify sediment during burial and lithification as it is brought into a subsurface environment of increased pressure, temperature, and changing pore-fluid compositions. These processes bring about a variety of changes in sediment characteristics, including compaction and cementation (porosity loss), dissolution of framework grains and cements (porosity gain), mineral recrystallization, and mineral replacement (see review in Boggs, 1992, ch. 9). Diagenesis can decrease the capacity of sediments and sedimentary rocks to transmit and store economically significant quantities of petroleum, natural gas, and ground water, as well as adversely affecting our ability to make reliable interpretations of provenance and depositional environments. On the other hand, some diagenetic processes have a positive economic impact. For example, dissolution can increase porosity of sediments and thus their ability to store hydrocarbons. Diagenesis is considered such an important topic among sedimentologists that hundreds of articles and more than 15 full-length books about siliciclastic diagenesis have been published in the English language alone (see, for example, Burley and Worden, 2003).

Diagenetic minerals and fabrics are commonly studied by using a variety of petrographic and geochemical techniques. Cathodoluminescence imaging is also making important contributions to our understanding of diagenetic processes and our ability to recognize diagenetic minerals and fabrics. It has been applied particularly to studies of porosity loss in sediments owing to compaction and cementation.

Diagenetic processes and products

Compaction and porosity loss

The porosity of newly deposited sediment can range from about 30 to 50 percent for sandy sediment and to more than 70 percent for carbonate sediment. Much of this porosity can be lost during diagenesis owing to a combination of compaction and cementation. Compaction alone can account for significant porosity reduction. Total compaction can be divided into two aspects: (1) mechanical compaction, which involves bulk volume reduction of sediments by grain rearrangement, plastic deformation of ductile grains, and breakage of brittle grains; and (2) chemical compaction, which involves bulk volume reduction as a result of pressure solution at grain contacts (Figure 5.1).

Effective study and interpretation of siliciclastic sedimentary rocks requires that we recognize fabrics and minerals that originate by diagenetic processes and differentiate them from original depositional features. In this regard, it is desirable to have a viable technique for estimating total compaction (mechanical and chemical compaction) and cementation in order to evaluate porosity and assess the relative importance of the various diagenetic events that act in concert to affect porosity (e.g., Houseknecht, 1991).

In an early (now-classic) paper, Sippel (1968) demonstrated that cathodoluminescence imaging is very useful in accomplishing many of

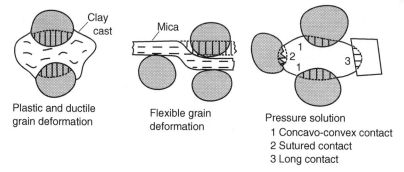

Fig. 5.1. Schematic representation of diagenetic textural features in sandstones that result from compaction. The hachured areas indicate rock volume lost by grain deformation and pressure solution. (From Wilson and McBride, 1988. Compaction and porosity evolution of Pliocene sandstones, Ventura Basin, California. *American Association of Petroleum Geologists* Bulletin, **72**, Fig. 10, p. 679. AAPG © 1988, whose permission is required for further use. Reproduced by permission of the AAPG.)

these objectives. Among other things, Sippel pointed out that weakly luminescent quartz overgrowths (cement) can be effectively differentiated from strongly luminescent detrital quartz cores (e.g., Figure 5.2). He also noted that many apparent sutured boundaries between quartz grains are the result of quartz cementation rather than pressure solution. Further, he described and interpreted quartz-filled fractures in detrital quartz grains, and discussed the possible relationship of such fractures to polycrystallinity in quartz.

Subsequent workers have built upon Sippel's pioneering work with CL imaging, which has evolved into an essential tool for study of various aspects of siliciclastic diagenesis. For example, Houseknecht (1987, 1991) discussed the process of making quantitative and qualitative estimates of compaction in quartz-rich sandstones on the basis of CL imaging. First, a photomosaic of CL images of a thin section is prepared and a "point count" of the photomosaic is made by placing some kind of grid over the photograph, analogous to doing a point count of a thin section with a petrographic microscope. The point count generates a quantitative estimate of the volume percent detrital quartz, quartz cement (overgrowths), and intergranular porosity. In a quartz-rich sandstone cemented with silica, CL imaging is essential for identifying quartz overgrowths that are not marked by "dust rims" of impurities or bubbles, such as those illustrated in Figure 5.2. The point-count data are then used to estimate the intergranular volume (the sum of intergranular porosity and intergranular cement). Paxton *et al.* (2002) suggest that intergranular volume also includes depositional (detrital) matrix. Intergranular volume provides an estimate of the total compaction, which represents both mechanical and chemical compaction.

Mechanical compaction

Mechanical compaction results from physical processes related primarily to burial pressures, although some aspects of mechanical compaction may be due to tectonic stresses. Thus, mechanical compaction involves bulk volume reduction of sediments as a result of rearrangement of grains (packing) as well as plastic deformation of ductile grains and breakage of brittle grains, as mentioned. Each of these processes can cause significant porosity reduction. For example, theoretical considerations show that even-sized spheres packed into the loosest arrangement (cubic packing) have a porosity of 47.6 percent. Simple rearrangement of the spheres to yield the tightest packing (rhombohedral packing) reduces the porosity to 26 percent (Graton and Fraser, 1935). Of course, the grains in natural

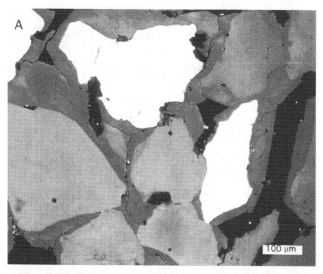

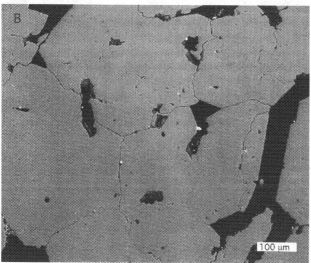

Fig. 5.2. Overgrowths (dark rims) on detrital quartz cores (bright), Garn Formation (Jurassic), North Sea continental shelf. (A) Cathodoluminescence image; (B) backscattered-electron (BSE) image. Note that the overgrowths cannot be differentiated from the detrital cores in the BSE image. (From Oelkers *et al.*, 1996 A petrographic and computational investigation of quartz cementation and porosity reduction in North Sea sandstones. *American Journal of Science*, **296**, Fig. 2B, p. 424 and Fig. 2D, p. 425. Reproduced by permission of the American Journal of Science.)

sands are neither spheres nor even-sized grains. Therefore, porosity loss owing to simple grain rearrangement cannot be predicted theoretically.

Plastic deformation

Plastic deformation of ductile lithic grains (Figure 5.1), such as micritic limestone clasts or shale or phyllite clasts, can substantially reduce intergranular volume in lithic-rich sandstones. Such deformed lithic grains may be difficult to differentiate from cement or matrix by using a petrographic microscope; however, they may be distinguishable by using CL, particularly if high-resolution color-CL images are available (Houseknecht, 1991).

Brittle deformation

At depths generally above those at which pressure solution commonly takes place, brittle fracturing of grains may be important. Fracturing reduces intergranular volume because grain fragments move into inter-granular spaces and adjacent grains can move into a closer packing arrangement (Houseknecht, 1991). Brittle deformation, identified on the basis of CL characteristics, has been documented by several investigators (e.g., Dickinson and Milliken, 1995; Laubach, 1997; Milliken and Lau-bach, 2000; Makowitz and Milliken, 2003). As discussed in Chapter 4, healed fractures are common in source-rock quartz grains, particularly in plutonic quartz. To evaluate correctly the importance of brittle fracturing to the overall process of compaction, investigators must be able to dif-ferentiate between healed fractures that originated in the source rocks (inherited fractures) and fractures that formed during diagenesis owing to compaction and subsequent precipitation of secondary quartz. Also, fractures created by grinding during thin-section preparation must be identified.

Laubach (1997) recognized three categories of quartz-filled (healed) microfractures that may be present in the quartz grains of sandstones that have undergone diagenesis, and which may be identified in CL images. Category I microfractures include those that cut across grain boundaries and cements without having a consistent relation to grain centers or grain contacts (e.g., Figure 5.3); they are common in well-cemented sandstones. Fractures that can be traced unambiguously across two or more adjacent grains are clearly post-depositional. Category II microfractures are more equivocal. They have curved to straight traces arranged in simple to complex intersecting patterns; the fracture traces curve moderately to strongly in radiating arrays associated with grain contacts, grain

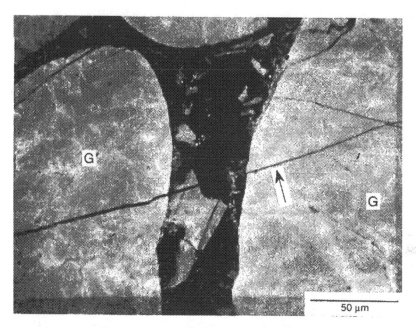

Fig. 5.3. Cathodoluminescence image showing a Category I microfracture (arrow) that extends across two adjoining quartz grains (G). Travis Peak Formation (Cretaceous), east Texas. Burial depth 2809 m. (From Laubach 1997. A method to detect natural fracture strike in sandstones. *American Association of Petroleum Geologists Bulletin*, **81**, Fig. 2a, p. 607. AAPG © 1997. Reproduced by permission of the AAPG whose permission is required for further use.

interpenetrations, or stylolites (e.g., Figure 5.4). Makowitz and Milliken (2003) further suggest that many Category II fractures are wedge-shaped with the widest fracture aperture at grain contacts. Expansion, or inflation, of individual grains by numerous fractures may create an exploded appearance in CL images, and extreme crushing may also occur along grain–grain contacts (Figure 5.5).

Category III microfractures are inherited fractures. According to Laubach (1997), the most obvious inherited fractures in CL images are wide fractures that end abruptly with blunt terminations at grain boundaries. He also suggests that inherited fracture fills may luminesce more brightly than the cement in post-depositional fractures. The stated criteria for distinguishing between Category II and Category III (inherited) fractures seem somewhat weak and subject to misinterpretation. For example, can healed microfractures such as those shown in the plutonic quartz grains in Figures 4.5 and 4.6 (Chapter 4) be unequivocally

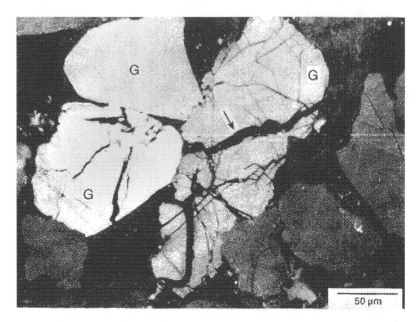

Fig. 5.4. Cathodoluminescence image of Category II microfractures in quartz grains (G), Canyon Sandstone (Permian), west Texas. Fractures radiate from grain contact (arrow). Burial depth 1820 m. (From S. E. Laubach, 1997. A method to detect natural fracture strike in sandstones. *American Association of Petroleum Geologists Bulletin*, **81**, Fig. 3b, p. 608. AAPG © 1997. Reproduced by permission of the AAPG whose permission is required for further use.

differentiated from quartz-filled, post-depositional fractures? The potential for error in interpreting inherited versus post-depositional fractures has relevance to both compaction studies and provenance analysis.

On the basis of CL imaging, Makowitz and Milliken (2003) reported that grain fracturing and subsequent cementation of fractures by quartz can occur at various depths before, during, and after normal quartz overgrowth precipitation. They suggest that failure to recognize authigenic quartz in fractures as cement rather than grain volume (i.e., by using light microscopy instead of CL) can lead to inaccurate estimates of inter-granular volume and loss of porosity by compaction and cementation.

The presence of fractures within quartz grains has implications in addition to their inferred effect on compaction. Fractures may be a major cause of polycrystallinity in detrital quartz grains. Milliken and Laubach (2000) reported that the relationship of polycrystallinity to intragranular fractures renders the use of polycrystallinity questionable for provenance

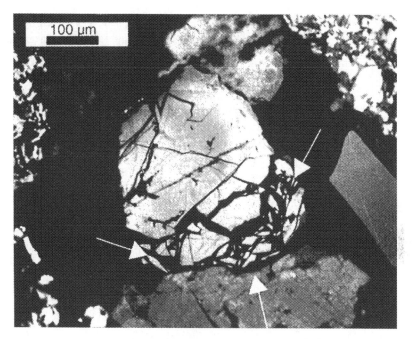

Fig. 5.5. Cathodoluminescence image showing extensive grain crushing (arrows) owing to brittle deformation. Frio Formation (Tertiary), Texas Gulf Coast. Burial depth ~ 3500 m. (From A. Makowitz and K. L. Milliken, 2003. Quantification of brittle deformation in burial compaction, Frio and Mount Simon Formation sandstones. *Journal of Sedimentary Research*, **73**, Fig. 3C, p. 1009. Reproduced by permission of SEPM.)

determination; polycrystallinity has been suggested by some previous workers (e.g., Young, 1976) to have provenance significance.

Chemical compaction (pressure solution)

Chemical compaction refers to the process whereby solution in sedimentary rocks occurs preferentially at the contact surfaces of grains where the external pressure exceeds the hydraulic pressure of the interstitial fluid; increased pressure leads to an increase in solubility at grain contacts. Solution of grains along a contact can result in interpenetration of one grain into another along what is referred to as a **sutured contact**. Pressure solution on a scale larger than individual grains can cause formation of dissolution seams called **stylolites**, which are particularly common in carbonate rocks. As indicated below in the discussion of grain contacts, CL can be useful for differentiating pressure-solution fabrics from cements.

Extensive pressure solution results in loss of porosity and thinning of beds. Stone and Siever (1996) report that porosity loss in quartzose sandstones caused by mechanical compaction and pressure solution occurs mainly at burial depths less than about 2 km because the combined effects of compaction, pressure solution, and a small amount of quartz cement produce stable grain-packing arrangements. According to these authors, porosity loss at greater depths is primarily the result of quartz cementation. Worden and Burley (2003), on the other hand, suggest that some porosity loss owing to compaction can continue to depths of at least 5 km.

Compaction and grain contacts

In a much-cited paper, Taylor (1950) described four types of contacts between grain that can be observed in thin sections: **tangential contacts**, or point contacts; **long contacts**, appearing as a straight line in the plane of a thin section; **concavo–convex contacts**, appearing as a curved line in the plane of a thin section; and **sutured contacts**, caused by mutual stylolitic interpenetration of two or more grains (Figure 5.6). In very loosely packed fabrics, some grains may not make contact with other grains in the plane of the thin section and are referred to as "floating grains". Contact types are related to both particle shape and packing. Tangential contacts appear only in loosely packed sediment or sedimentary rocks, whereas concavo–convex and sutured contacts occur in rocks that have undergone considerable compaction during burial. Increased compaction results in a general progression from tangential to long, concavo–convex, and sutured contacts. As pointed out by Sippel (1968) and Houseknecht (1991), CL petrography is extremely useful in avoiding misinterpretation of contact types. For example, growth interference between adjacent quartz overgrowths (cement) can easily be mistaken for grain contacts indicative of intergranular pressure solution (concavo–convex and sutured), unless CL imaging is used to identify the authigenic overgrowths.

Cathodoluminescence imaging can also resolve ambiguities in grain-contact types that arise owing to brittle fracturing. Milliken and Laubach (2000) report that a significant number of concavo–convex and sutured contacts are associated with a prominent degree of brittle deformation. Cathodoluminescence images reveal that many apparent sutured boundaries actually arise from spatial redistribution of grain fragments rather than by volume loss. Such grains show a pronounced increase in degree of brecciation immediately adjacent to the region of maximum

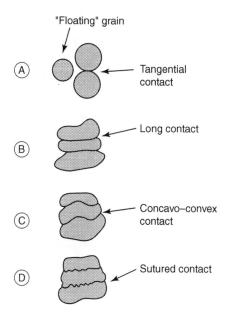

Fig. 5.6. Diagrammatic illustration of principal kinds of grain contacts: (A) tangential; (B) long; (C) concavo–convex; (D) sutured. (Based on Taylor, 1950. From Boggs, S., Jr., 2001. *Principles of Sedimentology and Stratigraphy*, 3rd edn., © 2001, p. 86. Reprinted by permission of Pearson Education, Inc, Upper Saddle River, NJ.)

curvature of the grain contact (Figure 5.7). In this case, spatial redistribution of fine-size crushed fragments in the indented grain is thus responsible for grain interpenetration, not pressure solution. As illustrated in Figure 5.7, conventional light microscopy provides no hint that the concavo–convex boundary was actually generated as a result of brittle fracture.

Houseknecht (1991) also described a method for estimating the amount of pressure solution that has occurred in sandstones by using CL petrography. Some quartz-rich sandstones may display very closely packed fabrics and a dominance of sutured grain contacts (e.g., Figure 5.8). Such fabrics indicate that the sandstones have undergone severe chemical compaction and loss of silica owing to pressure solution. In Houseknecht's technique, visible grain boundaries (in CL images) are projected across adjacent grains along contacts where pressure solution appears to have taken place, resulting in a lens-shaped area (termed overlap quartz) that represents portions of grains that are inferred to have been dissolved by pressure solution. Determining the relative

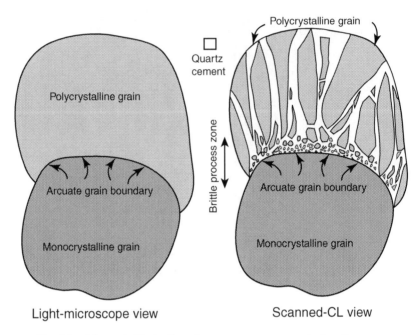

Light-microscope view   Scanned-CL view

Fig. 5.7. Schematic illustration of a concavo–convex contact as it appears
in a light-microscope view and a SEM–CL view. The light-microscope view
shows an arcuate grain boundary that is strongly suggestive of compaction
through pressure solution and volume loss from the upper polycrystalline
grain. Scanned-CL microscopy reveals that polycrystallinity in the upper
grain arises from brittle deformation and that interpenetration is due to
spatial redistribution of grain fragments and not pressure solution. (After
Milliken and Laubach, 2000. Brittle deformation in sandstone diagenesis as
revealed by scannned cathodoluminescence imaging with application to
characterization of fractured reservoirs. In Pagel, M., V. Barbin, P. Blanc,
and D. Ohnenstetter eds., *Cathodoluminescence in Geosciences*, Berlin,
Springer Verlag, Fig. 4, p. 231. Reproduced with kind permission of Springer
Science and Business Media.)

volume of these overlaps provides at least a semiquantitative estimate of
the volume of detrital quartz dissolved by intergranular solution.

Cementation

Cementation is a major chemical diagenetic process in siliciclastic sedimentary rocks, particularly in sandstones. Cementation can begin during
early stages of diagenesis, referred to as **eogenesis**, shortly after deposition of sediment. It may continue through deep burial (**mesogenesis**) and
may even occur under some conditions of late-stage diagenesis after uplift

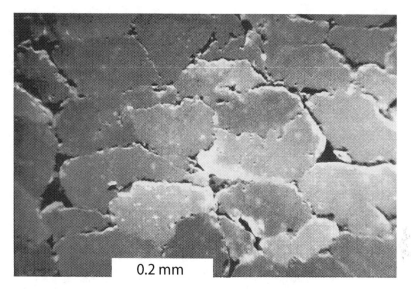

Fig. 5.8. Cathodoluminescence micrographs showing closely packed fabric, with a predominance of sutured contacts, owing to pressure solution in the Bromide Sandstone (Ordovician, Oklahoma). (From Houseknecht, 1988. Intergranular pressure solution in four quartzose sandstones. *Journal of Sedimentary Petrology*, **58**, Fig. 4D, p. 233. Reproduced by permission of SEPM.)

of sedimentary rocks into a near-surface environment (**telogenesis**). Cementation plays a major role in reducing the porosity of sedimentary rocks, and it can also affect compaction of sediments. For example, early-formed cements that fill pore spaces of a sandstone or shale before significant compaction tend to inhibit further compaction of the sediment.

A variety of minerals may occur as cements in siliciclastic sedimentary rocks. These cements include silica minerals (quartz, chalcedony, opal), carbonate minerals (calcite, aragonite, dolomite, siderite, ankerite), K-feldspars, albite, hematite, goethite (limonite), pyrite, gypsum, anhydrite, barite, chlorite and other clay minerals, zeolites, tourmaline, and zircon (Boggs, 1992, p. 378). Carbonate, silica, and clay-mineral cements are by far the most important, although zeolites may be abundant in some volcaniclastic sediments.

Quartz cements

Quartz is a common cementing material in sandstones, particularly in quartz-rich sandstones (quartz arenites). Quartz cementation has been discussed in a host of research papers, as well as in special monographs (e.g., *Quartz Cementation in Sandstones*, edited by Worden and Morad, 2000). As mentioned in preceding discussion, quartz cement commonly occurs as an overgrowth on detrital quartz grains. Overgrowths are in optical continuity with the detrital cores of the quartz grains; thus, they may be difficult to recognize by optical microscopy. They can commonly be easily differentiated from detrital cores in CL images by their less intense luminescence (e.g., Figure 5.2).

Some overgrowths are reported to contain CL growth banding (e.g., Walker and Burley, 1991; Lyon *et al.*, 2000; Goldstein and Rossi, 2002). The growth zones in overgrowths, which are commonly on the scale of tens of microns in width, presumably reflect changes in physico-chemical conditions during the time in which the overgrowths precipitated (e.g., Perny *et al.*, 1992). On the other hand, Goldstein and Rossi (2002) report SEM–CL zoning in quartz overgrowths from the Jurassic Khatatba Formation of Egypt, which they suggest formed through a multistage process that involved recrystallization of earlier precipitates (coatings) of less stable silica phases such as opaline silica and chalcedony. Some overgrowths display very complex intergrowth of zones. For example, Hogg *et al.* (1992) described quartz overgrowths in Brent Group (Jurassic) sandstones from the UK North Sea that contain as many as six CL zones (Figure 5.9). They suggested that quartz cementation occurred in several pulses, interrupted by periods of non-deposition, and that quartz cementation is a flux-controlled process. Apparently, in this case, hot, silica-bearing fluids from greater depths entered the sands periodically along a fault.

Because quartz overgrowths are abundant in many sandstones and because the overgrowths are comparatively easy to recognize in CL images, numerous published studies have included CL images of quartz overgrowths on detrital quartz grains; however, not all silica cement in sandstones is precipitated as overgrowths. Silica can also be precipitated into interstitial space among grains (without forming overgrowths) as opal, chalcedony, or microquartz (chert). Opal (an isotropic mineral) is a comparatively rare cement in sandstones except in some volcaniclastic successions where the opal is derived by alteration of volcanic glass. Chalcedony (also not a common cement) has a fibrous to feathery texture, and microquartz (chert) has an equigranular texture (e.g., Boggs, 1992, pp. 132–3). Microquartz (chert) cement is much less common then quartz overgrowth cement, but is present in some lithic arenites (e.g., Dapples, 1979) and even in some quartz arenites (e.g., Hoholick, 1984). Microquartz (chert) cement can occur both as an isopachous rim on

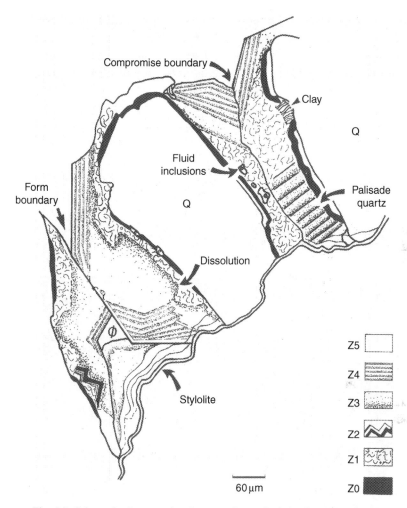

Fig. 5.9. Schematic diagram showing complex cathodoluminescence zoning in quartz overgrowths in Brent Group (Jurassic) sandstones from the UK North Sea. Zone ZO is a thin, non-luminescent zone adjacent to the detrital quartz (Q) that may contain some clay minerals, Z1 contains trapped fluid inclusions and may contain perpendicular CL texture (palisade quartz), Z2 forms quartz crystallites with oscillatory subzones, Z3 is relatively thick and homogeneous, Z4 contains oscillatory subzones that grade from bright CL to dark CL, and Z5 cuts across Z4 and is thin, bright (CL), and euhedral. (From A. J. C. Hogg E. Sellier, and A. J. Jourdan, 1992, Cathodoluminescence of quartz cements in Brent Group sandstones, Alwyn South, UK North Sea. In Morton *et al.*, eds, *Geology of the Brent Group*, Geological Society Special Publication, No. 61, Fig. 9, p. 433. Reproduced by permission of The Geological Society, London.)

quartz grains and as a mosaic of chert deposited within pore spaces (e.g., Figure 5.10). Microquartz cement commonly displays **drusy** texture (elongated crystals oriented normal to pore walls line the walls and project toward pore interiors). As discussed in preceding sections, precipitation of quartz as a fracture-filling cement is also very common and such fractures have been studied successfully by using CL (e.g., Milliken and Laubach, 2000). Repeated fracturing and cementation can create complex "crack–seal" microstructures (Ramsey, 1980).

Microquartz probably precipitates initially as opal-CT (crystobalite disordered by interlayered tridymite lattices), which is metastable and subsequently transforms to microcrystalline quartz (chert), e.g. Pettijohn *et al.* (1973, p. 427). Cathodoluminescence images of microquartz (chert) cement, as it appears in grayscale CL images, are not necessarily superior to petrographic images for identifying microquartz cement; however, CL

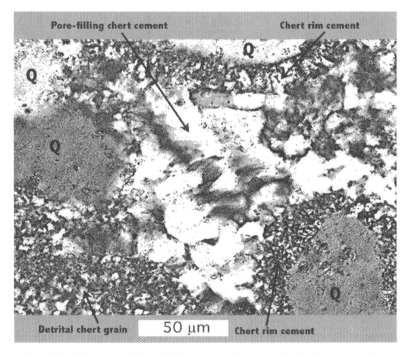

Fig. 5.10. Petrographic (optical) image of microquartz (chert) cement filling pore space among four quartz grains (Q) and one detrital chert grain (lower left corner). Note isopachous chert-cement rims around some quartz grains. The sample is from a chert-cemented sandstone interbed in the Jefferson City Dolomite (early Ordovician), Missouri. (Sample provided by Western Minerals, Inc.)

can, in some cases, provide significant additional insight into the process of cementation by microquartz. Figure 5.11A is a CL image of the same grains shown in the petrographic photograph in Figure 5.10. Note that at least three generations of chert cement are visible in the CL image. The BSE image (Figure 5.11B) also shows some of these generations. As far as we know, the principles of cement stratigraphy that are commonly applied to study of successive generations of carbonate cements in carbonate rocks (discussed in Chapter 6) have not been applied to study of chert cement in sandstones; we are not aware of any published studies in which CL was used to study chert cement. Nonetheless, the CL patterns displayed in Figure 5.11A indicate progressive filling of pore space with successive generations of silica cement, which appear similar in geometric pattern to successive generations of carbonate cement in carbonate rocks. Generation 1 cement in Figure 5.11A (bright CL), which forms iso-pachous rims on some quartz grains, precipitated first (oldest cement). Generation 2 cement (dull CL) precipitated next, and generation 3 cement (mottled, bright CL) precipitated last (youngest cement). Differences in CL intensity of these three generations of chert cement indicate changes in trace-element (activator-ion) concentrations in pore waters through time that reflect the diagenetic history of the sandstones. Because our work on chert-cemented sandstones is in a very preliminary stage, we have not yet identified the CL-activator ions in the different generations of cement. Thus, we are not at this time able to speculate on why activator-ion concentrations changed during precipitation of the different cement generations. We hope to expand research on chert-cemented sandstones in the future.

Cathodoluminescence color may also provide additional help in identification and evaluation of chert cements. Quartz precipitated from an aqueous fluid commonly shows weak luminescence, but it can display a variety of CL colors, including blue, green, yellow, red bottle-green, brown, and pink (e.g., Ramseyer *et al.*, 1988). These colors tend to be very transient (disappear rapidly after 20–30 seconds of electron irradiation).

As mentioned, quartz cementation has an important influence on sediment compaction and porosity loss. Some investigators have recently used CL analysis, in conjunction with other techniques, to study different phases of quartz cementation and relate them to burial history of sediments. For example, a detailed study of the diagenetic succession in Upper Paleozoic Haushi Group sandstones in Oman (Hartmann *et al.*, 2000) revealed that dissolution of aluminosilicates, pressure solution, and

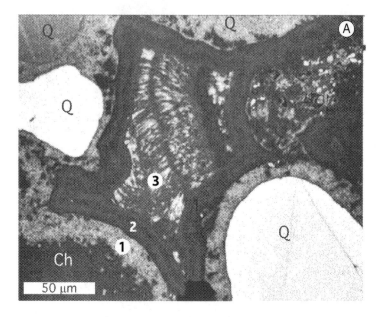

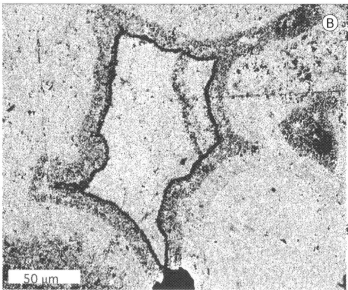

Fig. 5.11. (A) Cathodoluminescence, and (B) BSE images of approximately the same area of the chert-cemented sandstone shown in Figure 5.10. Q = quartz; Ch = detrital chert; 1, 2, and 3 indicate successive generations of chert cement (1 – oldest; 3 – youngest). Additional generations of chert cement may possibly be present.

stylotization were major silica-providing processes at different burial depths, leading to different phases of silica cementation. Other studies (e.g., Sedat, 1992; cited in Richter *et al.*, 2003) show that CL color of quartz cement can range from blue to brown, depending upon precipitation temperature and thus burial stage (eogenetic, mesogenetic, telogenetic).

Although most CL studies of diagenetic quartz have been applied to sand-size quartz, Schieber, *et al.* (2000) used CL imaging to study silt-size quartz in Devonian black shales of the eastern United States. Much of the quartz silt in these shales is nonluminescent, which suggests that it was not deposited as detrital quartz but instead formed authigenically. The authors conclude that this nonluminescent quartz silt originated during early diagenesis by precipitation of silica into algal cysts and other pore spaces. Clearly, CL petrography played an essential role in reaching this conclusion.

Feldspar cements

Feldspar cements occur primarily as overgrowths on detrital feldspar grains, on either K-feldspars or albite. Authigenic feldspar is most common in feldspathic and volcaniclastic sandstones but may be present also in some quartz arenites and lithic arenites. Feldspar overgrowths form both during eogenesis and some stages of mesogenesis. Feldspar overgrowths can generally be recognized in CL images by their less-intense luminescence compared to that of the detrital feldspar cores (e.g., Figure 5.12); however, overgrowths commonly emit some luminescence. Both Walker and Burley (1991) and Götze *et al.* (2000) report observing white and brown CL colors from orthoclase overgrowths and low-temperature albite in sandstones, and Richter *et al.* (2002) report dark-olive luminescence in authigenic albite.

Carbonate cements

Carbonate cements occur in all types of sandstones as well as in many shales. They appear to be most typical of quartz-rich sandstones (quartz arenites), but they are also quite common in many feldspathic arenites and may be present in some lithic arenites. Carbonate cements may occur as uniform pore fillings over large areas of a rock unit, or their distribution can be very patchy. Well-cemented zones can grade to poorly cemented zones over distances of a few meters. Carbonate cement can also occur as lenses or stringers or as concretions. Carbonate concretions are particularly common features of some shales. Calcite is the principal

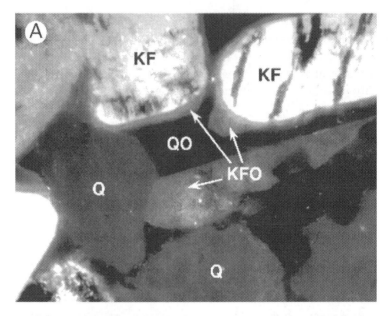

Fig. 5.12. (A) Cathodoluminescence image of microcline overgrowths (KFO) on detrital feldspars (KF); Q = quartz; QO = quartz overgrowth. (B) Optical image (crossed nicols) of the same grains. Triassic "Buntsandstein" of southwestern Germany. (From Richter *et al.*, 2003. Progress in application of cathodoluminescence (CL) in sedimentary petrology. *Mineralogy and Petrology*, 79, Plate 2, Fig. **g** and **h**, p. 131. Reproduced by permission of Springer-Verlag. Photographs courtesy of Thomas Götte.)

carbonate cement in siliciclastic sediments. Dolomite, ankerite (ferroan dolomite), siderite, and aragonite occur also; however, aragonite exists primarily in very young rocks. Carbonate cements may fill pore spaces among detrital grains with a mosaic of fine crystals, or a pore space may be filled with a single large crystal. In fact, a single crystal may grow large enough to surround several detrital grains to produce a poikilotopic texture.

Carbonate cements can commonly be readily identified in polarized petrographic (crossed nicols) images; therefore, CL imaging is not as important in identifying carbonate cements as it is in identifying quartz overgrowths. The CL characteristics of carbonate minerals are a function mainly of the relative abundance of $Mn^{2+}$and trivalent rare-earth ions, which are activators of extrinsic CL, and Fe^{2+}, which is a quencher of CL (e.g., Marshall, 1988; Machel, 2000). Cathodoluminescence colors in carbonate minerals can range from blue through red, depending upon the relative abundances of activator and quencher ions, which can vary in all of the carbonate minerals (discussed subsequently in Chapter 6). There-fore, it may be difficult to differentiate among (identify) different car-bonate-mineral cements on the basis of CL color; CL color can commonly be used effectively to distinguish between carbonate cements and detrital silicate framework grains.

Many carbonate cements display zoning (e.g., Figure 5.13), which arises from differences in relative abundances of activator and quencher ions. These differences reflect variations in characteristics of formation (pore) waters at various times during diagenesis; therefore, they are important in developing an understanding of burial history and have possible use in correlation. The study of zoning in carbonate cements has come to be known as **cement stratigraphy**, as mentioned. This kind of study has been applied primarily to cements in carbonate rocks. Cement stratigraphy is discussed in depth in Chapter 6 and is not considered further here. Although carbonate cements are abundant in many sand-stones (and shales), it appears that CL imaging has been little used in their study, in sharp contrast to study of quartz cements in sandstones.

Clay-mineral cements

Clay minerals make up between 30 and 60 percent of the total minerals in shales. They are common as detrital constituents; however, one kind of detrital clay mineral may alter to another during diageneis, e.g., smectite commonly alters to illite. Thus, shales may contain significant amounts of authigenic clay minerals. Clay minerals are also present in the pore spaces

Fig. 5.13. Cathodoluminescence image of calcium-rich, zoned dolomite, Burlington Formation (Mississippian), Iowa. The differences in luminescent intensity result primarily from small differences in Fe and Mn contents. (From Reeder and Prosky, 1986. Compositional sector zoning in dolomite. *Journal of Sedimentary Petrology*, **56**, Fig. 2b, p. 239. Reproduced by permission of SEPM.)

of many sandstones, particularly in lithic arenites and volcaniclastic sandstones, as matrix or cement. Although some clay matrix in sandstones may be detrital, most probably originated by mechanical infiltration of fine-size material into pore spaces after deposition or by chemical alteration of framework minerals such as feldspars during diagenesis. Clay minerals formed by chemical alteration are considered to be authigenic matrix. They may also be precipitated directly from pore waters as cement, without involving alteration of precursor framework grains. Clay-mineral cements precipitated into open pore space can commonly be recognized by the presence of drusy texture; however, authigenic clay matrix may be difficult to differentiate from detrital clay matrix (see discussion in Boggs, 1992, p. 150).

The principal clay minerals in sedimentary rocks are kaolinite, smectite, illite, and chlorite, although serpentine, pyrophyllite, talc, and micas may also be regarded as clay minerals. Many of these clay minerals can also form during diagenesis. For example, kaolinite can form diagenetically at low temperatures ($< \sim 25\,^{\circ}C$) during eogenesis or telogenesis, as well as at higher diagenetic temperatures. Diagenetic Na-smectites form particularly by alteration of volcanic glass at temperatures ranging from < 25–150 °C. Both illite and chlorite are generated during mesogenesis at temperatures ranging between ∼ 55 and > 200 °C.

Comparatively few studies have been made of the CL characteristics of clay minerals. Marshall (1988, p. 50) noted that kaolinite-group minerals, serpentine, pyrophyllite, talc, some micas, and Fe-free chlorite all show some luminescence. The first comprehensive study of the luminescent properties of clay minerals from different localities and different origins (sedimentary, diagenetic, hydrothermal) appears to have been made by Götze *et al.* (2002). This study showed that all dioctahedral clay minerals of the serpentine–kaolinite group exhibit blue CL, whereas other clay minerals – e.g., serpentine, talc, montmorillonite (smectite), illite – do not. Thus, the blue-CL color of kaolinite-group clay minerals allows them to be rapidly distinguished from other common clay minerals in sandstones, such as smectite and illite. Furthermore, different members of the kaolinite group may be distinguished from each other on the basis of different time-dependent behavior of CL-emission intensity during electron irradiation. Kaolinite and halloysite show a transient lumines-cence intensity, which rapidly decreases within ∼60 seconds of electron bombardment, whereas the intensity of blue CL from dickite, nacrite, and pyrophyllite increases during irradiation. Thus, different pore-filling clay-mineral cements may be distinguished by CL imaging. In some cases, they may even be quantified in combination with computer-assisted image analysis (e.g., Götze and Magnus, 1997).

Apparently, most reported studies of the CL colors of clay minerals have been performed by using cathodoluminescence microscopes, which have limited capability for magnification. The potential exists for higher-resolution CL studies of clay minerals by using SEM–CL, employing instruments equipped with optical filters. Also, live color-CL imaging in a scanning electron microscope is now possible by using a CL detector (ChromaCL) newly introduced to the market by Gatan, as mentioned in Chapter 3.

Mineral replacement and neomorphism

Mineral **replacement** involves the gradual dissolution of one mineral and essentially concurrent precipitation of another mineral in its place, e.g., replacement of feldspars by carbonate minerals. Replacement processes commonly produce distinctive replacement textures, such as cross-cutting relationships, illogical mineral composition (pseudomorphs), and bite-like embayments of the guest (replacing) mineral into the host (replaced) mineral that are called caries texture (Boggs, 1992, Table 9.5). Replacement textures can generally be recognized in both petrographic (optical) images and backscattered SEM images. We have found little indication in the published literature that CL imaging has been used to study replacement textures. Clearly, however, the potential to utilize CL petrography (especially SEM–CL) to study replacement features is there. Differences in luminescence characteristics of silicate minerals such as quartz and feldspars, clay minerals, carbonate minerals, sulfides, sulfates, phosphates, etc. (Marshall, 1988, ch. 4) can surely be utilized to differentiate guest minerals from host minerals, particularly in partially replaced minerals.

Recrystallization refers to changes in size or shape of crystals of a given mineral, without accompanying change in chemical composition or mineralogy. **Inversion** is a more complex kind of recrystallization in which one polymorph of a mineral changes to another polymorph without change in chemical composition, e.g., inversion of aragonite ($CaCO_3$) to calcite ($CaCO_3$). Folk (1965) suggested the term **neomorphism** as an inclusive term that incorporates both inversion and recrystallization. Neomorphism is an extremely important process in carbonate rocks – a topic to which we return in Chapter 6. Neomorphism is relatively much less important in siliciclastic sedimentary rocks. Examples of neomorphism in siliciclastic rocks include extensive recrystallization in plagioclase crystals and transformation of opal cements to microquartz (Boggs, 1992, p. 401). Cathodoluminescence imaging has apparently been little used to study neomorphism in siliciclastic sedimentary rocks; however, again, the potential is there.

6

Luminescence characteristics and diagenesis of carbonate sedimentary rocks

Introduction

Carbonates are an important group of rocks that make up nearly one-quarter of all rocks in the sedimentary record. They differ fundamentally from siliciclastic sedimentary rocks both in composition and in depositional origin. Siliciclastic sedimentary rocks are composed of constituents that originated outside the depositional basin and were transported as solids into the basin. By contrast, carbonate rocks are composed of intrabasinal sediments, which were deposited *in situ* by precipitation of calcium carbonate through inorganic and biochemical processes. They have little or no provenance significance, unlike siliciclastic sedimentary rocks; however, the textures and structures that characterize carbonate rocks have considerable significance with respect to interpreting depositional conditions (e.g., water depth, water energy, biological activity). Thus, petrographic study of carbonate rocks has customarily focused on observation of carbonate grains (including fossils), textures, and structures that are important to interpretation of depositional conditions.

Carbonate rocks also have considerable economic significance, particularly as reservoir rocks for petroleum. Their importance as reservoir rocks depends upon the degree to which original (depositional) porosity (40–80 percent) is preserved during diagenesis. Porosity is lost owing to physical compaction, chemical compaction (pressure solution), and cementation. Diagenesis can also bring about pervasive changes in mineral composition (e.g., alteration of aragonite to calcite or dolomite and alteration of calcite to dolomite). The role of cathodoluminescence imaging in the study of carbonate rocks has been confined mainly to study of carbonate diagenesis, although it plays some role in identification of carbonate grains, textures, and structures. Cathodoluminescence has found special application to evaluation of zoning in carbonate cements and interpretation of zoning with regard to the burial history of carbonate sediments and the nature of diagenetic pore waters.

The common minerals that make up carbonate sedimentary rocks fall into three main groups:(1) the **calcite group** (calcite, magnesite, rhodochrosite, siderite, smithsonite); (2) the **dolomite group** (dolomite, ankerite); and (3) the **aragonite group** (aragonite, cerussite, strontionite, witherite). Calcite ($CaCO_3$), aragonite ($CaCO_3$), and dolomite [$CaMg(CO_3)_2$] are the principal minerals of geologic interest and the principal carbonate minerals considered herein.

Luminescence centers and CL characteristics of carbonate minerals

The luminescence characteristics of the carbonate minerals are controlled primarily by the relative abundances of manganese, rare-earth elements (REEs), and iron. The Mn^{2+} ion and trivalent REE ions appear to be the most important activator ions of extrinsic CL, whereas Fe^{2+} is the principal quencher (e.g., Marshall, 1988; Machel, 2000, Richter *et al.*, 2003). Cathodoluminescence spectra of some natural carbonates are shown in Figure 6.1, after Richter *et al.* (2003). Figure 6.1 A–C are calcite spectra, Figure 6.1D is an aragonite spectrum, and Figure 6.1E, F are dolomite spectra.

Calcite

Luminescence centers

The CL characteristics of calcite are summarized in Table 6.1. The most common (extrinsic) luminescence colors of calcite are orange-yellow, yellow-orange, and orange (e.g., Marshall, 1988; Habermann *et al.*, 2000a). Cathodoluminescence colors may also include blue and, less commonly, green, which may be related to intrinsic (structural) defects. The Mn^{2+} ion appears to be the most important activator in calcite, although some REE ions may also be activators (e.g., Machel, 2000). Spectral measurements indicate that the major Mn^{2+}-activated calcite emission peak occurs at $\sim$605–620 nm (e.g., Habermann *et al.*, 2000a); however, a peak at 590 nm has also been reported (cited in Marshall, 1988). Also, less intense, more poorly defined (intrinsic?) Mn^{2+} peaks may occur at wavelengths between $\sim$400–700 nm (see Figures 6.1 A, B).

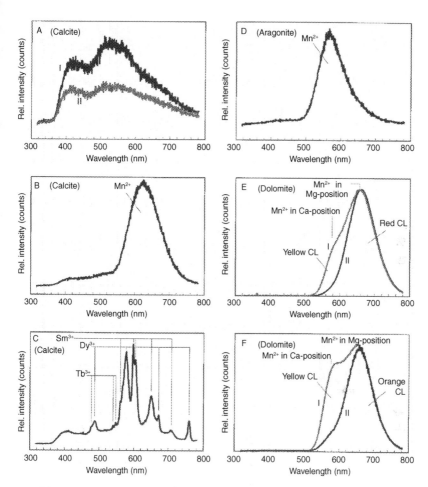

Fig. 6.1. CL spectra of some natural carbonates from Germany and Austria. (A) Holocene *calcite* with intrinsic CL (stalagmite, Bavaria), I, clear calcite; II, pigmented calcite. (B) Subrecent *calcite* (stalagmite, North Rhine-Westphalia) with Mn-activated CL. (C) Middle Holocene *calcite* (stalagmite, North Rhine-Westphalia) with REE-activated CL. (D) *Aragonite* (bivalve, Rhine near Bonn), with Mn-activated CL. (E) Permian *dolomite* (sulfatic evaporite, northwest Hessen), I, yellow; II, red zoned, Mn-activated luminescence. (F) Hydrothermal dolomite (Austria) with yellow (I, two maxima at 575 and 656 nm) and red (II, one maximum at 656 nm) Mn-activated luminescence. [After Richter *et al.*, 2003, Progress in application of cathodoluminescence (CL) in sedimentary petrology. *Mineralogy and Petrology*, **79**, Fig. 7, p. 141. Reproduced by permission of Springer-Verlag.]

Table 6.1. *Peak positions in the CL spectrum and CL colors of calcite, aragonite, and dolomite*

Mineral	Wavelength (nm)	Activator ion(s)	CL colors	References
Calcite	590, (605–620)	Mn^{2+}	Orange-yellow, yellow-orange, orange, violet(?)	Habermann et al. (2000a), Richter et al. (2003), Marshall (1988)
	~500, 545, 580, 600, 650, 680, 710, 760	REE (particularly Sm^{3+}, Dy^{3+}, Tb^{3+})	Sm^{3+}-activated CL may be same as Mn^{2+}-activated CL; Tb^{3+} activates green CL and Dy^{3+} activates cream-white CL	Machel (2000), Habermann et al. (1996)
Aragonite	~(400–700)	Intrinsic	Blue (weak)	Habermann et al. (2000a)
	540, 560	Mn^{2+}	Yellow-green, green	Marshall (1988), Richter et al. (2003)
Dolomite	Two main peaks (570–583), (649–659)	Mn^{2+}	Yellow, red	Richter et al. (2003)

Note: REE-activated CL is outshone by much stronger Mn^{2+}-activated CL, if Mn_{2+} is present at >10ppm (Habermann et al., 1996).

Emission peaks activated by REEs (particularly, Dy^{3+}, Sm^{3+}, and Tb^{3+}) may be present at several wavelengths (e.g., Figure 6.1C).

The Fe^{2+} ion is considered a major quencher of extrinsic CL in carbonate minerals (e.g., Machel, 2000; Richter *et al.*, 2003). Therefore, the intensity of CL emissions in calcite reflects the abundance of Fe^{2+} as well as that of Mn^{2+}. Marshall (1988, p. 83) suggests that luminescence intensity appears to be controlled by the Fe/Mn ratio, not by the absolute concentrations of either cation; however, Habermann *et al.* (1998, 2000a) report that Mn^{2+} concentrations as low as 1 ppm can activate CL emissions in Fe-poor calcites and that a linear correlation exists between Mn^{2+} concentrations and CL intensity at Mn^{2+} values ranging from 19–1000 ppm. These authors suggest that Fe^{2+} contents up to 2000 ppm have no influence on this relationship. Manganese$^{(2+)}$-activated CL is quenched at >3000–4000 ppm Fe^{2+} and the degree of quenching is controlled by the Fe^{2+} and Mn^{2+} content. These observations appear to disagree somewhat with those of Budd *et al.* (2000); these authors report that 25 ppm Mn are required to initiate visible CL and that Mn is the sole control on CL intensity when Mn and Fe contents are very low (< 100 ppm). Their data also show that 100 ppm Fe is the minimum for quenching CL when Mn is below 200 ppm. Further, their research indicates that Pb and Zn concentrations in tens of parts per million neither activate nor sensitize CL.

A question still remains as to whether or not the CL intensity of carbonate minerals is controlled entirely by the concentration levels of Mn^{2+} (+ some REE?) and Fe^{2+}. Some workers have suggested a more complex relationship that involves other factors, discussed below. Haberman *et al.* (2000a) indicate that self-quenching becomes important at high concentration levels of Mn^{2+} (> 1000 ppm). Therefore, CL intensity appears to be a complex function of Mn^{2+} activation and both Mn^{2+} self-quenching and Fe^{2+} quenching. Although Mn^{2+} and Fe^{2+} are the principal ions that affect CL emissions in carbonates, numerous factors influence the partitioning of these ions into carbonate minerals. These factors or processes include activity coefficients, activity of calcium, various chemical species in solution, temperature, crystal surface structures, crystal growth rates, pH, and redox reactions (see Machel and Burton, 1991 and Machel, 2000, for a more complete listing and explanation of these factors). According to Machel (2000), these processes fall into four groups: changes in redox potential (which may be difficult to evaluate), closed-system element partitioning (trace-element variations in diagenetic fluids are governed by the degree of "openness" of the system), organic matter maturation, and clay-mineral diagenesis (organic matter and clay-mineral diagenesis are possible sources and sinks of many trace elements, including known activators). Exotic sources, such as sources of hydrothermal fluids, may also contribute trace elements.

Cathodoluminescence zoning

Carbonate minerals are typically characterized by CL zoning. Zoning reflects either changing growth conditions or variations in growth mechanisms. Two types of zoning are recognized: (1) **concentric zoning**, in which the compositional interface of the zoning pattern coincides with or is parallel to the growth interface existing at the time of growth; and (2) **sectoral zoning**, in which the compositional interface does not coincide with and is not parallel to the growth interface (Reeder, 1991). An example of concentric zoning superimposed on sectoral zoning is shown in Figure 6.2.

Concentric zoning is generally considered to reflect changing bulk-fluid properties, which in turn reflect fluid chemistry, temperature, pressure, and possibly redox conditions. Bulk-fluid properties may not, however, be entirely responsible for all zoning in all cases. The partitioning of trace elements between a crystal (calcite) and the fluid in which it grows is influenced by the crystal growth rate, as mentioned. The partitioning of Mn^{2+} and Fe^{2+} into calcite, for example, decreases with growth rate and is also possibly affected by temperature (Dromgoole and Walter, 1990). Thus, some zoning in calcite may be caused by variations in rates of trace-element (e.g., Mn^{2+} and Fe^{2+}) partitioning into calcite.

Aragonite

Aragonite (orthorhombic) does not have exactly the same CL properties as calcite (rhombohedral), even though it has the same major-element chemical composition. The luminescence colors of aragonite are yellow-green and green (e.g., Marshall, 1988; Richter *et al.*, 2003). The maximum Mn^{2+}-activated emission peak is at $\sim560\,nm$ (Fig. 6.1D). According to Barbin (2000) the difference in CL color of calcite and aragonite is due to changes in crystal-field parameters. The free Mn^{2+} energy levels are modified and the emission occurs as a result of the transition from the $^4T_{1g}$ excited state to the $^6A_{1g}$ ground state. The Mn content of aragonite can apparently be quite variable. Marshall (1988) reports that some aragonite contains no detectable Mn and is nonluminescent.

Fig. 6.2. Concentric zoning superimposed on sectoral zoning in a calcite cement crystal. Note that overall CL intensity changes abruptly at the growth sector boundaries, some of which are indicated by arrows. (After Reeder and Grams, 1987. Sector zoning in calcite cement crystals: implications for trace element distributions in carbonates. *Geochimica et cosmochimica acta*, **51**, Fig. 1b, p. 188. Copyright 1997, with permission from Elsevier.)

Dolomite

Dolomite may display either red or yellow luminescence colors; however, somewhat different peak positions of Mn^{2+}-activated dolomite CL spectra have been reported. El Ali *et al.* (1993) indicated that dolomite shows two main peaks at 578 nm and 655 nm. The CL peak at 578 nm is due to Mn^{2+} in the Ca site and the peak at 655 nm to Mn^{2+} in the Mg position. These authors also suggested that the CL color of dolomite may

have environmental significance. Specifically, they reported that evaporitic dolomite from the Jurassic Mano Formation has homogeneous
red emission with two peaks at 583 nm and 659 nm. Recent evaporitic
dolomites from Mexico and dolomites from Oman show yellow emission
with bands at 577 nm and 650 nm. Sedimentary non-evaporitic dolomites
generally luminesce yellow-orange with bands at 570 nm and 649 nm,
whereas, hydrothermal dolomites show various luminescence colors
ranging from orange to red and with spectral bands at 575 nm and
656 nm.

Richter *et al.* (2003) state that the CL spectra of yellow to red luminescent dolomites are generally dominated by the emission band at
656 nm, but have distinct asymmetric intensity profiles at lower wavelengths, which is caused by the two overlapping bands (see Figure 6.1E,
F). They indicate that the reason for varying Mn^{2+} distributions in early
diagenetic dolomites, as well as in dolomites of higher-grade diagenetic to
metamorphic environments, are a point of discussion owing to incon-
sistent results reported by various investigators (e.g., El Ali *et al.*, 1993).
They further suggest that yellow or red luminescence colors in dolomites
are not just an effect of different environments; growth kinetics may also
be important.

Many dolomite crystals are characterized by compositional zoning,
which is generally displayed much more distinctly in CL images than in
petrographic (optical) images. Cathodoluminescence zoning can be both
concentric and sectoral (see Figure 5.13). Differences in intensity of CL
emitted from different zones reflect mainly variations in Mn^{2+} and Fe^{2+}
concentration levels in the crystals.

Luminescence of marine carbonates

Conventional wisdom has generally dictated that marine carbonates,
particularly young marine carbonates, either do not luminesce or exhibit
very dull luminescence, (e.g., Machel, 1985; Major, 1991). This point of
view has been applied to both biogenic and inorganic marine carbonates;
however, Barbin (2000) reported that most recent biogenic carbonates
exhibit luminescence to some degree. Presumably the concentration of
divalent Mn in normal seawater is too low, when partitioned into marine
carbonate minerals, to activate visible luminescence in abiotic carbonates
in the presence of typical levels of divalent Fe partitioned from normal
seawater. Apparently, Mn in the divalent state is rare in seawater,

compared to Mn^{3+}, in the well-oxygenated shelf environments in which most carbonate sediments are deposited. Thus, high levels of divalent Mn are to be expected mainly in reducing environments.

Reported exceptions to this general dictum of nonluminescent marine carbonates have been cited, which may be related to oxygen-depleted pore waters (a few centimeters below the sediment–water interface) that contain divalent Mn (e.g., Major, 1991). Nonetheless, most young abiotic marine carbonates apparently do not display intense cathodolumines-cence. Some workers have reasoned from this assumption that brightly luminescent marine-carbonate minerals must, therefore, have undergone recrystallization, although Major (1991) urges caution in making this interpretation.

Applications

Carbonate petrology

The major components of limestones are carbonate grains (clasts, ooids, pellets, fossils), sparry calcite cement, and microcrystalline calcite (micrite) (e.g., Scholle and Ulmer-Scholle, 2003). The classification and environmental interpretation of limestones depends upon recognition and identification of these components in thin sections. Cath-odoluminescence imaging is a useful supplement to petrographic (optical) study for this purpose. For example, Miller (1988) used CL imaging to illustrate the internal features of microfossils and ooids (Figure 6.3), as well as cements, and shows that these features may be revealed more clearly in CL images than in petrographic (optical) images. Adams and MacKenzie (1998) present several color plates in which both optical images and CL images of the same views are shown. Details of cement textures, shapes of dolomite crystals, zoning in dolomite crystals, and other textural features are displayed to significant advantage in the CL images. In their impressive new color guide to the petrology of carbonate rocks, Scholle and Ulmer-Scholle (2003) briefly discuss the use of CL in carbonate petrography and present color-CL plates showing dolomite crystals and cements. Richter *et al.* (2003) offer several beautiful color plates illustrating CL features such as echinoid spines (Figure 6.4), bra-chiopod shells, and zoned dolomite crystals. Marshall (1988) also includes several color-CL plates illustrating such features as crinoidal grainstones and details of bryozoan shells and ooids.

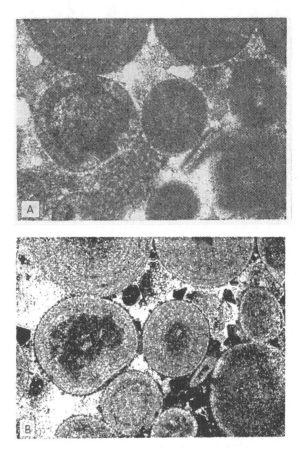

Fig. 6.3. (A) Optical (crossed nicols) image of ooids cemented with calcite cement (white); and (B) CL image of the same grains showing that details of ooid fabrics are revealed much more clearly. (From Miller, 1988. Cathodo-luminescence microscopy. In Tucker, M. ed., *Techniques in Sedimentology*, Oxford, Blackwell Scientific Publications, Fig. 6.4e, f, p. 184. Reproduced by permission.)

On the whole, however, CL imaging has not been used extensively to study the depositional components and textures of carbonate rocks. By far, the greatest application of CL imaging to carbonate rocks has been to analysis of carbonate diagenesis. The remainder of this chapter deals with that topic.

Carbonate diagenesis

The diagenesis of carbonate sediments may involve organic processes (e.g., boring and shell breakage by organisms), physical processes (e.g.,

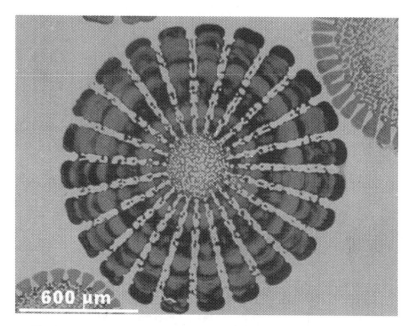

Fig. 6.4. Cathodoluminescence image of a recent echinoid spine (cross-section view), Hegoland, Germany. [From Richter *et al.*, 2003, Progress in application of cathodoluminescence (CL) in sedimentary petrology. *Mineralogy and Petrology*, **79**, Plate 4a, p. 148. Reproduced by permission of Springer-Verlag. Photograph courtesy of Thomas Götte.]

compaction and bed thinning, grain breakage), and chemical processes (e.g., dissolution, cementation, neomorphism). **Eogenesis** begins on the seafloor with organic reworking of sediments, early cementation, neomorphism, and possibly dissolution, depending upon environmental conditions. The sediments may subsequently undergo a phase of meteoric diagenesis if a shallow depositional basin fills with sediment to or above sea level or if sea level falls, exposing previously formed carbonates. Diagenesis in the meteoric zone may include cementation, dissolution, alteration of Mg-calcite to calcite, and neomorphism of aragonite to calcite. **Mesogenesis** begins when subsidence eventually brings carbonate sediments into the deep-burial environment where physical processes (compaction and pressure solution) take place and where further chemical changes occur (cementation, neomorphism, mineral replacement, dissolution). Diagenetic effects can continue with late-stage uplift of deeply buried sediment and exposure in the meteoric zone (**telogenesis**).

Although CL imaging may have some applications to study of diagenetic features stemming from physical processes, almost all published CL study of carbonate diagenesis has focused on observation and analysis of features generated by chemically mediated processes. The bulk of these studies has been directed toward analysis of the CL characteristics of cements, particularly zoned cements. The CL study of these cements and the genetic and diagenetic interpretations drawn on the basis of differences in the CL characteristics of different zones in cements is referred to as **cement stratigraphy**. Cathodoluminescence petrography has applications also to study of neomorphic processes such as recrystallization of dolomite.

Cementation and cement stratigraphy

The cementation history of carbonate sediments may be quite complex given that carbonate cements can be precipitated in seafloor (depositional), meteoric, or deep-burial diagenetic environments. Carbonate cements may be composed of aragonite, calcite, Mg-calcite (>4% $MgCO_3$ in solid substitution), or dolomite and may exhibit a variety of textural forms depending upon the diagenetic environment (see review by Boggs, 1992, ch. 12). For example, seafloor cements may be fibrous and form isopachous rinds around carbonate grains; meteoric cements may include meniscus and pendant cements; and deep-burial cements may display bladed-prismatic texture, coarse mosaic texture, or syntaxial overgrowths on echinoderm fragments. This listing of cement types is greatly simplified and incomplete, but it indicates that carbonate cements can be quite variable in composition and texture. Furthermore, cements formed in one diagenetic environment may be added to and enlarged by precipitation of another (younger) generation of cement in a different diagenetic environment.

As pointed out by Machel (2000), the most widespread use of CL in carbonate studies is in cement stratigraphy by using zoned cements. During the progressive burial and diagenesis of a given carbonate deposit, one generation of cement may be deposited in the seafloor environment, another in the meteoric environment, and still another in the deep-burial environment. With careful study it may be possible to detect each of these generations of cement. For example, it may be possible within single pores or adjacent pores in a limestone to identify a marine seafloor cement (oldest), a meteoric cement (younger), and a deep-burial cement (youngest). This technique of studying successive generations of cement, referred to as cement stratigraphy as mentioned, was introduced by Evamy (1969). It was first applied in cathodoluminescence studies by Meyers (1974), who used the concept to establish lateral and chronological correlation of cements from the Mississippian rocks of New Mexico. The technique has subsequently been employed by many investigators as a tool for evaluating the diagenetic history of carbonate rocks in relation to the changing characteristics of diagenetic pore fluids during progressive burial.

In some samples, different generations of cements may be tentatively identified simply by petrographic analysis; for example, a cloudy seafloor cement (oldest) may be present together with a clear meteoric cement (youngest). Most cement stratigraphy requires more-detailed study that may involve staining, cathodoluminescence, ultraviolet microscopy, trace-element analysis, and fluid-inclusion studies. For example, staining with potassium ferrocyanide can reveal differences in ferrous iron content of calcite or dolomite. These differences may reflect changes in oxidizing conditions of the precipitating fluid, such as a change from the meteoric to the deep-burial environment. Further, trace-element analysis using instruments such as the ion microprobe can reveal the concentration levels of Mn and Fe, etc. By far, however, the most heavily used tool in cement stratigraphy is cathodoluminescence.

Cement stratigraphy is based primarily on study of concentrically zoned cements (Meyers, 1991). It is commonly accepted that concentric zones nearer to the substrate (pore wall) are older than those farther from the substrate, as illustrated in Figure 6.5. Five major CL zones (generations of cement) are visible in this CL image of cement-filled pores in Mississippian limestones from New Mexico; 1 is the oldest generation, 5 the youngest. Figure 6.6A is another example that shows a cement-filled cavity in a pisoid Devonian limestone from Western Australia (Wallace *et al.*,1991) that is filled by an older fibrous meniscus cement and a younger clear calcite cement. Three generations of clear cement are depicted with great clarity in the CL image (Figure 6.6B). Note that the clear calcite cement in this image grades from nonluminescent (oldest) to brightly luminescent to dully luminescent (youngest). Wallace *et al.* (1991) state that as many as five major zones may be present in the clear calcite in some pores of this limestone.

Numerous other workers have described zoned cements and reported anywhere from two (e.g., Meyers, 1991) to six (e.g., Bourque *et al*, 2001) cement zones/generations. (Evaluating the number of cement generations in published work can be difficult because different workers refer to successively deposited cements as generations, sequences, zones, sub-zones, stages, or substages; or they simply number the cements 1, 2, 3, etc.) A commonly reported succession of major CL generations in

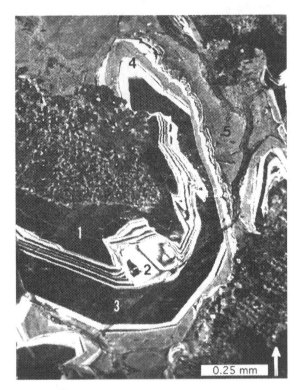

Fig. 6.5. Zoned carbonate cements in crinoidal packstones of the Lake Valley Formation (Mississippian), New Mexico. Five zones are shown in this CL image from 1 (oldest) through 5 (youngest). [After Meyers, 1974. Carbonate cement stratigraphy of the Lake Valley Formation (Mississippian), Sacramento Mountains, New Mexico. *Journal of Sedimentary Petrology*, **44**, Fig. 6C, p. 844. Reproduced by permission of SEPM (Society for Sedimentary Geology)].

cements is nonluminescence (oldest) to bright luminescence to dull luminescence (e.g., Grover and Read, 1983; Dorobek, 1987; Wallace *et al.*, 1991). Essentially the reverse succession, bright (oldest) to dull to nonluminescent, has also been reported (e.g., Lee and Harwood, 1989). Richter *et al.* (2003) report that a common sequence of zones observed worldwide consists of four generations of calcite cement, as revealed by CL color. (I) The cement sequence starts with a calcite cement generation with patchy brown-orange luminescence. (II) The second generation of cement shows an intrinsic blue luminescence. (III) The third generation displays medium-to dark-orange luminescence, commonly with sectoral zoning. (IV) The fourth generation, not present in all cases, consists of an

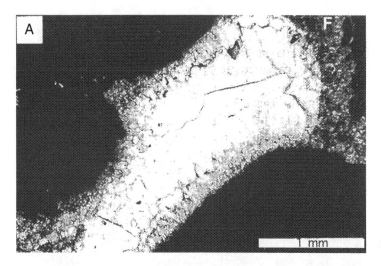

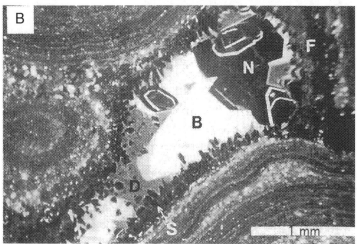

Fig. 6.6. (A) Thin-section photomicrograph (plane-polarized light) of a cement-filled cavity in a pisoid limestone, filled by a fibrous meniscus cement (F) and a clear cement. (B) Cathodoluminescence image of the same area in (A). Nonluminescent, elongated crystals (S) rim the cavity. The clear, equant cement consists of an early nonluminescent cement (N) overgrown by a brightly luminescent cement (B) that, in turn, is overgrown by a dully luminescent cement (D). Pillara Limestone (Devonian), Western Australia. (From Wallace *et al.*, 1991. Burial diagenesis in the Upper Devonian reef complexes of the Geikie Gorge Region, Canning Basin, Western Australia. *American Association of Petroleum Geologists Bulletin*, **75**, Fig. 6A,B, p. 1023. AAPG © 1991. Reprinted by permission of the AAPG whose permission is required for further use.)

intrinsically luminescing cement with fine orange luminescing growth zones.

Interpreting the significance of these zoning trends requires an understanding of the factors that affect CL intensity in carbonates. The causes for CL zoning in carbonate cements and the interpretation of zoning sequences have been points of discussion among geologists for some time. A general consensus appears to be emerging that the CL intensity (brightness) of the cement zones is mainly a function of the concentrations of Mn^{2+}-activator ions and Fe^{2+}-quencher ions, although a number of factors have been suggested to affect the partitioning of trace elements into carbonate, and thus the content of activator and quencher ions in carbonates, as mentioned.

On the basis of empirical studies, first reported by Machel *et al.* (1991), Machel (2000) constructed the diagram shown in Figure 6.7 to illustrate the relationship between the concentration of divalent Mn and Fe in calcite and dolomite and the intensity of CL emissions. This diagram indicates that the intensity of CL emissions is a complex function primarily of activation by Mn^{2+} and both Fe^{2+} quenching and Mn^{2+} self-quenching at high Mn^{2+} concentrations.

The concentration levels of divalent Mn and Fe in cements are, in turn, related to the redox conditions of the depositional and diagenetic environments, as well as several other factors (e.g., Machel, 2000). Seawater in the shallow-shelf environment, where most carbonate sediment is deposited, is commonly well oxygenated. Therefore Mn is present in marine cements mainly as Mn^{4+}, not Mn^{2+}, and Fe is present as Fe^{3+} rather than Fe^{2+}. Marine cements, which tend to have a fibrous form, thus exhibit dull CL. The vadose (unsaturated) meteoric environment is also largely well oxygenated. Cements in the vadose zone are likely to be meniscus cements (precipitated from a curved meniscus of water held between grain contacts) or pendant or gravitational cements (precipitated from drops of water that cling to the bottoms of grains) and can thus be differentiated from marine cements on the basis of form. In any case, seafloor cements and vadose-zone cements are likely to contain low levels of divalent Mn and both exhibit low-intensity CL. Subsidence of sediment into the phreatic zone of the meteoric environment brings sediments into an environment where sediment pores are filled (saturated) with water and redox conditions are reducing, at least in the lower part of the zone. Cements in this zone, which may be blocky, isopachous, or epitaxial, are likely to contain much higher levels of divalent Mn and display bright luminescence if divalent Fe concentration is low (if, for example,

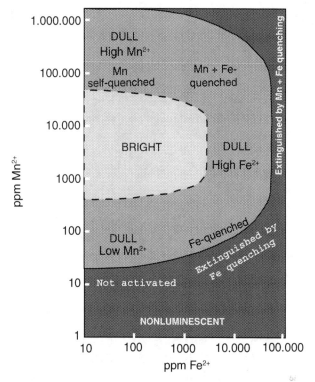

Fig. 6.7. Relation of CL intensity to Mn^{2+} and Fe^{2+} concentration in calcite and dolomite. Nonluminescence is caused by low Mn^{2+} concentrations (lower left) and extinction of CL is due to Fe^{2+} quenching as well as self-quenching of Mn^{2+} (right and upper right). Cathodoluminescence is bright at moderately high concentrations of Mn^{2+} if Fe^{2+} concentration is below about 2000 ppm (Habermann *et al.*, 2000a). Dull luminescence at very high concentrations of Mn^{2+} is due to Mn self-quenching (upper left) and Mn+Fe quenching (upper center). (Modified slightly from Machel *et al.*, 1991. Causes and emission of luminescence in calcite and dolomite. In Barker C. E. and O. C. Kopp, eds., *Luminescence Microscopy and Spectroscopy: Qualitative and Quantitative Applications*, SEPM Short Course 25, Fig. 12B, p. 20. Reproduced by permission of SEPM.)

Fe^{2+} is locked up in interactions with organic matter and sulfates). Continued subsidence brings sediments into the deep-burial environment where cements may be bladed prismatic, coarse mosaic, or epitaxial. Reducing conditions prevail in the deep-burial environment (which favors the presence of Mn^{2+} and thus bright CL); however, sufficient divalent Fe may be available to bring about some quenching of CL, which causes these cements typically to exhibit dull CL.

Typical observed successions of CL zones in cement have commonly been interpreted to signify a succession of burial events that involve a stage of cementation in the oxidizing meteoric environment (non-luminescent), followed by precipitation of cement in the reducing phreatic environment (bright CL), and finally precipitation in the deep-burial environment (dull CL). This succession might involve subsequent progressively deeper burial of a given unit of rock; however, it has also been suggested that cements with different CL intensities could be precipitated synchronously by meteoric ground waters that became progressively more reducing as they flowed downdip through paleoaquifers from the meteoric zone to deep-burial zones (e.g., Dorobek, 1987). Telogenesis introduces a new element into cement zoning, because it brings deeply buried sediments, in which cements were precipitated under reducing conditions (bright or dull CL), back into an oxidizing meteoric environment. Thus, the last major sequence of cement formed during telogenesis is nonluminescent cement precipitated in the vadose zone (e.g., Bruckschen *et al.*, 1992; Lee and Harwood, 1989).

Finally, not all nonluminescent to bright to dull luminescent packages can be linked to an episode of meteoric diagenesis (Emery and Marshall, 1989). For example, Stow and Miller (1984) describe similar sequences of zones from deep-sea sediments that could not have been exposed to meteoric conditions. According to these authors the package of zoned cement begins with an initial zone that is poorly luminescent, followed by a thin bright zone, progressing to a zone of medium luminescence with subzones, passing into a uniformly luminescent zone, which forms the bulk of the cement in the grainstones. In this case, fluctuations in Eh (oxidation potential) of pore waters and content of divalent Mn and Fe (necessary to generate poorly luminescent to bright to dull CL) must be attributed to (unknown?) processes other than meteoric diagenesis.

Correlation by cement stratigraphy

The basic principles of cement stratigraphy were enunciated through the pioneering work of Evamy (1969) and Meyers (1974), as mentioned. Meyers used CL to study zoned cements and identify distinctive zones that could be used for correlation. According to Meyers (1974, 1978), correlation of CL zones in cements can be achieved between adjacent samples within single measured sections, and between adjacent measured sections. He correlated cement zones in Mississippian rocks of New Mexico over several hundred meters of stratigraphic interval and laterally over about 16 km in a north–south direction. Subsequently, many other workers have used cement stratigraphy to correlate between samples similar to the way that lithostratigraphy allows correlation from one locality to another (e.g., Grover and Read, 1983; Dorobek, 1987; Goldstein, 1988; Kaufman *et al.*, 1988; Bruckschen *et al.*, 1992). Zonal correlation may be time synchronous and record specific events such as a rapid fall in sea level over a region, which exposes carbonate sediments to meteoric waters, or it may be diachronous, i.e., time transgressive (e.g., Goldstein, 1991).

Goldstein (1991) describes the practical aspects of zonal correlation. The first step is to identify and describe zonal patterns at the most detailed level observable in a single thin section, utilizing such features as CL color, brightness, thickness of zone, and grouping of zones. The second step involves correlation of cement zones at the thin-section scale. That is, attempts must be made to match the zonal pattern from one pore to the next within a thin section. If this is not possible, it may be possible to lump cement zones in a way that results in a unique sequence of cement-zone packages that can be correlated around the thin section. Once correlation has been established within a single thin section, the correlation is extended vertically in the stratigraphic section by making side-by-side comparison of samples. This procedure is illustrated in Figure 6.8. The last step is to correlate major cement zonal patterns (established vertically in the section) laterally between stratigraphic sections. This step requires the availability of (laterally) closely spaced samples that can be compared side by side.

Correlation of cement zones laterally does not necessarily mean that a chronologic (time-stratigraphic) correlation has been established, because events that generate cement zones may be time transgressive (e.g., slow, gradual fall of sea level). Goldstein (1991) suggests that the most useful technique to establish time significance of cementation events is to observe cross-cutting relationships that have clear time significance. Such features might include: cross-cutting relationships with paleosol features developed below surfaces of subaerial exposure that can be dated; cross-cutting relationships between cement zones and paleokarst features of known age; the relation of cement zones to compaction fabrics such as stylolites; and cross-cutting relationships between cement zones and fracture sets whose ages can be related to development of a particular fault or fold of known age. Age might also be established by correlating a cement zone to an erosional unconformity surface formed during an episode of subaerial exposure. For example, Carlson, *et al.* (2003) used cement stratigraphy and cross-cutting relationships to identify as many

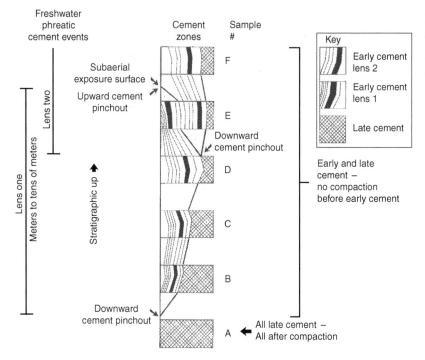

Fig. 6.8. Schematic illustration of hypothetical vertical correlation of cement zones in a stratigraphic section. Each letter designates the sequence of cement zones in an individual sample. The late cement postdates compaction whereas the early cements predate compaction. Note the downward pinchout of zones between samples E and D, and between B and A that represents the deepest stratigraphic extent of early cements. Note the upward pinchout of early cement zones between samples E and F that represents the stratigraphically highest extent of some of the early cement zones. Example based on the Pennsylvanian Holder Formation. (From Goldstein, 1991. Practical aspects of cement stratigraphy with illustrations from Pennsylvanian limestone and sandstone, New Mexico and Kansas. In Barker C. E. and O. C. Kopp, eds., *Luminescence Microscopy and Spectroscopy: Qualitative and Quantitative Applications*, SEPM Short Course 25, Fig. 2, p. 125. Reproduced by permission of SEPM.)

12 surfaces of subaerial exposure in Carboniferous limestones in the Brooks Range, Alaska.

In spite of fairly widespread use of cement stratigraphy in correlation, some geologists remain skeptical of its application. For example, Searl (1988) reports in a study of Carboniferous rocks in south Wales that individual zones can only be traced with any certainty between adjacent pores and are commonly identifiable only in a few square millimeters of thin sections. Further, cement sequences with apparently identical cathodoluminescence appearance are often demonstrably of different ages. That is, cements with dissimilar patterns may be of the same age and cements with similar patterns may have different ages. Machel and Burton (1991) and Machel (2000) further elaborate on these difficulties. Machel (2000) suggests that additional data, such as isotope ratios and fluid-inclusion homogenization temperatures and freezing-point depressions, should be used in connection with CL for unambiguous interpretation. In particular, additional data are necessary to make reasonable estimates of the physico-chemical characteristics of diagenetic pore waters. One common use of CL has been to identify marine components (cements or fossils) that are relatively unaltered by recrystallization in order to determine the isotopic composition of paleo-ocean waters (e.g., Popp *et al*, 1986; Lohmann and Walker, 1989; Tobin *et al.*, 1996).

Neomorphism in carbonates

As mentioned in Chapter 5, neomorphism refers to diagenetic changes that take place owing to polymorphic transformations (solution–reprecipitation) and/or recrystallization (Folk, 1965). Owing to the presence of relatively high levels of Mg in the modern world ocean, which inhibit crystallization of calcite, aragonite and Mg-calcite are the principal carbonate minerals precipitated in the modern ocean (see discussion in Boggs, 2001, p. 192). So-called "aragonite seas", which preferentially precipitated aragonite, also dominated the world ocean between about middle Mississippian and middle Jurassic time. Calcite precipitated preferentially at other times, when high rates of seafloor spreading increased removal of Mg^{2+} ions from seawater by absorption onto hot seafloor basalts (Sandberg, 1983; Stanley and Hardie, 1999).

Carbonate sediments are particularly prone to neomorphic changes during diagenesis. Aragonite is a metastable polymorph that inverts to calcite in the presence of water by a solution–reprecipitation process; Mg-calcite also tends to alter to (low-magnesian) calcite by dissolution–reprecipitation. The process of converting aragonite and Mg-calcite to calcite is commonly referred to as **calcitization**. Calcitization is a very important early diagenetic process (e.g., in the meteoric environment), which can alter the mineralogy of large volumes of carbonate sediment. In addition to changes in mineralogy, carbonate sediments may undergo extensive recrystallization; that is, change in crystal size and/or shape. For example, microcrystalline carbonate sediments (micrite) tend to recrystallize to much coarser, granular, or fibrous fabrics, and whole fossils and fossil fragments also commonly undergo recrystallization.

We found little evidence in the published literature that CL has been used as a tool to study calcitization, although CL spectroscopy could possibly be utilized to differentiate between aragonite and calcite. Cathodoluminescence has been employed to study some aspects of recrystallization in carbonate rocks. For example, Coniglio (1989) used CL to differentiate between equant calcite cement and equant neomorphic calcite (neospar). He reports that, compared to cement, neospar mosaics tend to have obvious curved intercrystalline boundaries (Figure 6.9) and lack enfacial junctions (a triple junction between three crystals where one of the angles is 180°). Cathodoluminescence imaging has also been used to resolve the origin of syntaxial rims on skeletal grains, which resemble syntaxial cements. On the basis of petrographic study, some workers have interpreted these rims as neomorphic replacements of micrite. Walkden and Berry (1984) used cathodoluminescence to restudy the origin of syntaxial rims in Lower Carboniferous limestones in Great Britain that were previously considered to be of neomorphic origin. On the basis of CL analysis, they concluded that the rims formed by passive cementation in solution voids that formed around echinoderm fragments during diagenesis. In another CL study of syntaxial rims in limestones, Maliva (1989) concluded that the rims formed by displacive crystal growth, which simply pushed aside the enclosing micrite.

Cathodoluminescence has been used much more extensively to study dolomite diagenesis. Dolomite is an enigmatic rock with respect to its origin. Dolomite formed in modern environments is primarily poorly ordered, Ca-rich "protodolomite," in contrast to well-ordered, stoichiometric dolomite in older rocks. Some dolomite may have formed initially as a primary precipitate (controversial); however, most dolomite likely formed by early replacement of a $CaCO_3$ precursor or subsequent replacement during deep burial. Pertinent models that have been proposed to explain early-formed dolomite are discussed in Boggs (2001, p. 197). Both the stoichiometry and ordering of dolomite appear to increase in a general way with increasing age. Presumably this increase indicates greater stabilization of dolomite owing to solution–reprecipitation and recrystallization phenomena. Thus, protodolomite apparently alters in time to stoichiometric dolomite.

Numerous investigators have utilized CL in some way to investigate dolomite diagenesis. Several studies have documented multistage alteration and recrystallization of dolomite by using cathodoluminescence and

Fig. 6.9. Plane-light photomicrograph (A) and cathodoluminescence image (B) of neospar in limestones of the Cow Head Group (Cambro-Ordovician), western Newfoundland, Canada. Note curved intercrystalline boundaries and lack of interfacial junctions. [From Coniglio, 1989. Neomorphism and cementation in ancient deep-water limestones, Cow Head Group (Cambro-Ordovician), western Newfoundland, Canada. *Sedimentary Geology*, **65**, Fig. 5a,b, p. 20. Copyright 1989, with permission from Elsevier.]

complimentary techniques such as trace-element and isotope analysis. For example, Reinhold (1998) describes multiple stages of dolomitization and dolomite recrystallization in Upper Jurassic shelf carbonates of southern Germany during shallow burial (Figure 6.10). An early-formed, Ca-rich, Sr-poor massive dolomite formed at shallow burial depth in Late Jurassic time. Two recrystallization phases occurred during further burial at elevated temperatures. Strong Sr enrichment of the second phase of recrystallized dolomite is ascribed to Sr-rich meteoric waters descending from overlying aragonite-bearing reef limestones or evaporite-bearing peritidal carbonates. Late-stage coarsely crystalline dolomite cements occur as vug and fracture fillings, and formed during burial. Note that the

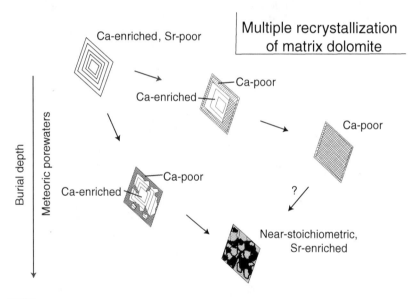

Matrix dolomite-A, bright zoned cathodoluminescence

Recrystallized matrix dolomite-B, homogeneous dull cathodoluminescence

Recrystallized matrix dolomite-C, clotted dull cathodoluminescence

Fig. 6.10. Schematic diagram showing cathodoluminescence patterns of selected matrix dolomite crystals summarizing diagenetic pathways during burial alteration and recrystallization. Successive recrystallization is indicated by changing CL patterns, Ca impoverishment, and Sr enrichment. (From Reinhold, C., 1998. Multiple episodes of dolomitization and dolomite recrystallization during shallow burial in Upper Jurassic shelf carbonates: eastern Swabian Alb, southern Germany. *Sedimentary Geology*, **121**, Fig. 11, p. 85. Copyright 1998, with permission from Elsevier.)

CL character of the dolomite changes from bright zoned in the early-formed dolomite to homogeneous dull to clotted dull with increasing burial and recrystallization.

A number of other applications of CL to study of dolomite diagenesis have been reported. For example, Nielsen *et al.* (1994) used CL, trace-element analysis, and O, C, and Sr isotope studies to document multiple-step recrystallization within massive dolomites from the Dinantian (Lower Carboniferous) of Belgium. Kupecz and Land (1994) used cathodoluminescence imaging to identify several generations of dolomite in Ellenburger Group dolomites (Lower Ordovician) of west Texas. Dark-brown-luminescent dolomite rhombs are overgrown by nonluminescent dolomite and dark-brown-luminescent dolomite. These early generations are cemented and replaced by orange-luminescent dolomite. The dolomites display covarient trends between increasing textural modifications (recrystallization) and increasing stoichiometry, decrease in trace-element concentrations, and depletion in ^{18}O.

Durocher and Al-Aasm (1997) used CL, petrographic, chemical, and isotopic studies to evaluate dolomitization in the Mississippian (Visean) Debolt Formation, northeastern British Columbia. They documented two early dolomitic phases, (1) early matrix dolomite and (2) pervasive dolomite, and two later generations, (3) coarse dolomite cement, and (4) pseudomorphic replacement of crinoids by dolomite. On the basis of CL petrography, stable-isotope studies, and fluid-inclusion analysis, Yoo *et al.* (2000) identified three stages of dolomitization, involving different fluids, which took place in Trenton and Black River limestones (Middle Ordovician) of northern Indiana. Gilhaus and Richter (2001) identified an early to late diagenetic dolomite succession (synsedimentary dolomite plus three stages of recrystallized dolomite) in an Upper Permian sabkha cycle from Hydra (Greece). They report that CL color in cement and replacing crystals changed from yellow-brown zoned to darkly zoned to faded red with increasing recrystallization. These examples are offered to illustrate that CL is being used as a routine tool in many studies of dolomite diagenesis. Numerous other examples could be cited.

7

Miscellaneous applications of cathodoluminescence to sedimentary rocks

Introduction

The most important applications of cathodoluminescence imaging to description and interpretation of sedimentary rocks are presented in modest detail in preceding chapters. Additional, less common, uses of CL in study of sedimentary rocks are discussed briefly in this chapter. These miscellaneous applications include CL study of recent and fossils shells (skeletal petrology), evaluation of the CL characteristics and significance of nonskeletal apatite, economic applications such as the use of CL to study petroleum reservoirs and sedimentary ore deposits, application to study of archeological materials, and use of CL to investigate characteristics of very old (Precambrian) sedimentary rocks.

Skeletal petrology

As mentioned in the discussion of carbonate petrology in Chapter 6, numerous workers have used CL imaging to study carbonate fossils because CL images commonly show skeletal outlines and internal details more clearly than do optical images. See, for example, Figure 6.4. cathodoluminescence can be particularly useful in identifying poorly preserved fossils. Barbin (1995, 2000) provides an extensive list of references dealing with application of CL to delineation of skeletal details and identification of fossils. Most CL studies of fossils have been applied to carbonate fossils; however, CL is also useful for studying skeletal material composed of apatite (francolite) and silica (e.g., Richter *et al.*, 2003).

Carbonate skeletons

In addition to its use in skeletal morphology and fossil identification, CL is being used to search for ontogenetic (development stages in growth of individual organisms) and paleoenvironmental records in skeletal organisms, as well as investigation of how CL patterns differ between biochemical and diagenetic CL emissions. Richter and Zinkernagel (1981) and Barbin (2000) have made particularly useful contributions in this area of research. More than 60 minerals have been identified in the carbonate skeletons of different organisms (e.g., Lowenstam and Weiner, 1989); however, only calcite, aragonite, and Mg-calcite are important skeletal-forming minerals. Some workers have suggested that biogenic carbonates do not luminesce, or luminesce very poorly (e.g., Major, 1991); however, recent work by Barbin (2000) demonstrates that most Recent biogenic carbonates do exhibit some luminescence. Cathodoluminescence emission is primarily orange in biogenic calcite, green to yellow in aragonite, and nonluminescent to blue in some Mg-calcite (i.e., in echinoderms). Cathodoluminescence may also be useful for identifying anomalous mineralization in biogenic skeletons, such as calcitic domains in aragonitic shells of organisms (Richter and Zinkernagel, 1981).

An open question remains with regard to variations in CL emissions in biogenic carbonates. Do variations in trace-element composition of biogenic carbonates reflect partitioning of trace elements from seawater owing to environmental factors (environmentally forced incorporation of Mn^{2+}) or to growth processes of organisms (metabolic-forced incorporation of Mn^{2+}). Environmentally and/or metabolic-forced incorporation commonly leads to intense growth zonation (orange zones in calcite and Mg-calcite, greenish-yellow zones in aragonite). Richter and Zinkernagel (1981) suggest that individual zones may not represent annual growth stages; a sequence of zones more likely represents annual growth stages. On the other hand, Barbin (2000) indicates that CL intensity is related to growth increments and, in bivalve shells, bands of maximum emission intensity are commonly correlated with annual growth lines and lower growth rates; further, CL emission increases with ontogeny. He also suggests that the amount of incorporation into biogenic carbonates is primarily controlled by metabolic activity; however, environmental factors such as water chemistry, temperature, and depth may influence the CL signal.

Diagenesis can likewise influence the CL signal. Richter and Zinkernagel (1981) point out that aragonitic skeletons lose their luminescence zonation during replacement by calcite, whereas Mg-calcite skeletons may keep part of it because their replacement preserves the original crystal fabric. Blotchy luminescence develops in Mg-calcitic particles during their adjustment to lower Mg-calcite by dissolution–precipitation processes in solutions with changing Mn/Fe ratios. Only in skeletons primarily composed of low Mg-calcite (e.g., brachiopod shells) can Mn-zonation be preserved. See also the review by Richter *et al.* (2003).

Phosphatic and siliceous skeletons

A few organisms have skeletal components composed of apatite (conodonts, lingulate brachiopods, vertebrate teeth and bones). The principal apatite mineral in these skeletal materials is a carbonate hydroxyl fluorapatite called **francolite** (general formula $Ca_5(PO_4, CO_3)_3(F, OH)$. The use of CL in the study of francolite skeletons is apparently in its infancy. Only a few published papers touch on this subject (e.g., Gaft *et al.*, 1997; Habermann *et al.*, 2000b; Richter *et al.*, 2003), and these papers deal mainly with the CL characteristics of conodonts. The most important activator elements in apatite are Mn^{2+} (green, broad emission band at 580 nm) and REE^{3+} ions (several narrow red, orange, violet, or pink bands).

Organisms that secrete siliceous skeletons include diatoms, radiolarians, and some sponges. These siliceous skeletons are composed of opal-A, which is amorphous SiO_2. Opal-A is metastable and crystallizes in time to opal-CT, an intermediate metastable phase between opal-A and quartz. Opal-CT is composed mainly of low-temperature cristobalite disordered by interlayered tridimite lattices. In turn, opal-CT is eventually converted into microquartz (chert, chalcodeny). Opal-A does not exhibit visible CL. Cristobalite and tridymite show blue CL with an emission maximum at 450 nm (Marshall, 1988); however, CL investigations of opal-CT have been inconclusive owing to its intimate intergrowth with other minerals, especially cryptocrystalline calcite (Richter *et al.*, 2003). According to Richter *et al.*, the visible CL color of biogenic chert is initially blue but changes to brownish-violet or brown with increasing irradiation, and CL emission at 650 nm increases.

Study of nonskeletal apatite

In addition to its presence in skeletal structures of some organisms, apatite is common as nonskeletal grains in some sedimentary rocks, and in many igneous, metamorphic, and hydrothermal rocks. The CL characteristics of apatite have been investigated by several workers. For example, Mariano (1988) reported that apatite in granitic rocks and pegmatites commonly has a strong yellow to yellow-orange fluorescence whereas apatite from carbonitites is blue and that from peralkaline syenite is pink-violet. Thus, the CL color of detrital apatite may have some provenance significance. Apatite that displays dark-green, bluish-green, and purple CL has been reported from some rare-metal deposits (Kempe and Götze, 2002). Color plates published by Waychunas (2002) and Kempe and Götze (2002) indicate that apatites can also exhibit strong growth zoning. Published CL spectra of apatite show that apatite CL can be activated by Mn^{2+}, with an emission peak at about 565 nm, as well as by several rare-earth elements with various peak positions (e.g., Gaft et al., 1997; Barbarand and Pagel, 2001; Kempe and Götze, 2002; Mariano, 1988; Mitchell et al., 1997; Waychunas, 2002).

Apatite is a relatively common mineral in sedimentary rocks. It occurs as detrital grains in sandstones and as primary precipitates and diagenetic minerals (e.g., nodules) in phosphorites, shales, and carbonate rocks. It is an abundant mineral in phosphorites, but CL has apparently been little used in study of sedimentary phospohorites. A few examples serve to indicate the potential of CL study of sedimentary apatite; however, relatively few workers have investigated the CL properties of sedimentary apatites in depth. In his review of apatite luminescence, Waychunas (2002) mentions that diagenetically produced sedimentary apatites generally show weak luminescence (possibly allowing it to be discriminated from detrital apatite?), although spectral analysis reveals the presence of REEs and Mn. Barbarand and Pagel (2001) describe the CL properties of apatite grains in drill cores from Upper Triassic sandstones, France. The apatite grains are composed of rounded cores and euhedral overgrowths. The yellow-CL cores are dominated by the Mn^{2+} band with only very small bands for the REEs (Gd^{3+}, Ce^{3+}, Dy^{3+}, Sm^{3+}). Spectra of the bright pink (CL) overgrowths are free of the Mn^{2+} band and have higher concentrations of Gd^{3+}, Ce^{3+}, Dy^{3+}, and Sm^{3+}. The authors suggest that the diagenetic fluid that formed the apatite rims was enriched in REEs compared with the original fluid from which the cores crystallized and also had a lower concentration of Mn^{2+}.

Lev et al. (1998) used CL, together with complementary techniques such as trace-element analysis and backscattered electron microscopy, to study the paragenetic sequence of mineralization in Ordovician black shales, Wales, UK. The observed paragenetic sequence was: pyrite formation → apatite in the form of rims and small grains → poorly crystalline carbonate → apatite nodules → monzanite replacing apatite and forming in the shale matrix → a second phase of carbonate replacing earlier-formed carbonate. Cathodoluminescence imaging allowed distinction between diagenetic apatite and other mineral phases. Lev *et al.* conclude that the formation of REE-rich apatite shows that REEs were redistributed and indicates that the initial depositional and provenance information recorded by the whole rock has been disturbed. They further suggest that paragenetic studies such as this are necessary to evaluate fully the usefulness of trace-element proxies susceptible to diagenetic redistribution.

Applications to sedimentary ore deposits

The principal application of cathodoluminescence petrography to evaluation of sedimentary ore deposits has been to study Mississippi Valley-type (MVT) ore deposits, although CL does have applications to other kinds of sedimentary ore deposits (e.g., study of carbonate fossils and cements in sedimentary ironstones, Hagni, 1986). The MVT deposits are lead and zinc concentrations that are hosted in carbonate sedimentary rocks. The ore minerals are galena (lead sulfide) and sphalerite (zinc sulfide), which are commonly associated with pyrite and marcasite and accessory minerals such as barite, gypsum, and fluorite. The host rocks for the ore minerals are limestones and dolomites; thus, dolomite and calcite are common gangue (non-ore) minerals. Cathodoluminescence can be useful in the search for ore and for tracing ore bodies from one district to another. Gangue minerals, particularily dolomite, are important in CL studies because they are widely distributed both laterally and vertically, and they can commonly be traced throughout a district even where the ore minerals are absent (Kopp, 1991, Kopp *et al.*, 1995). Mineralizing fluids that pass through carbonate rocks in a region affect the trace-element composition, and thus the CL characteristics, of gangue minerals associated with ore minerals. Distinctive CL zones may be recognized that represent different mineralizing events.

In an early paper dealing with this subject, Ebers and Kopp (1979) recognized six CL zones in dolomite, identified on the basis of CL color and intensity, which fill fractures in the MVT Mascot-Jefferson City District, Tennessee. The authors were able to correlate these CL zones through much of the mineralized district. Other early applications of CL to study of MVT deposits include Kopp *et al.* (1986) and Rowan (1986). Kopp *et al.* (1986) used CL microscopy to trace gangue carbonates (both dolomite and calcite) for more than 50 km in east Tennessee and more than 40 km in central Tennessee. On the basis of their study, they suggest that the MVT districts in east and central Tennessee may belong to a single district extending well over 200 km. Rowan (1986) used CL microscopy to study zoned hydrothermal, vug-lining dolomites in the Viburnum Trend lead district of southeast Missouri. He reported that CL microstratigraphies in the dolomites provide a time framework in which fluid-inclusion measurements can be tied, and suggests that CL stratigraphies defined in the Viburnum Trend may be traced in dolomite cements for a distance of up to 350 km to the southwest.

Keller *et al.* (2000) provide an example of a more recent application of CL to study of the migration of mineralizing fluids. These authors used CL microscopy together with stable-isotope and fluid-inclusion studies to evaluate fluid migration and associated diagenesis in the greater Reelfoot Rift region, Midcontinent, USA Cathodoluminescence petrography revealed nine distinct carbonate microstratigraphies in the Reelfoot Rift and on its adjacent carbonate platforms in Missouri (western platform) and Tennessee (eastern plaftorm). Geochemical data indicate that stratigraphically controlled fluid flow through carbonate aquifers in the Reelfoot Rift was restricted within individual fault blocks that had limited communication with adjacent carbonate platforms.

In another case study, Montañez (1997) used CL cement stratigraphy to delineate regional diagenetic and fluid migration events associated with Mississippi Valley-type mineralization in the southern Appalachians, USA. Multiple generations of replacement dolomite and CL-zoned dolomite and calcite cements occur in Lower Ordovician Upper Knox Formation carbonates. Zoned dolomite and calcite cements, and associated late-diagenetic replacement dolomite, paragenetically bracket MVT mineralization; further, the regional distribution of CL-zoned carbonate cements far exceeds the geographic and stratigraphic distribution of individual MVT deposits. Knox carbonate cements exhibit a complex cathodoluminescent stratigraphy that can be correlated regionally throughout mineralized and unmineralized regions. Montañez interpreted cement zonal correlations to have time significance and to record regionally extensive diagenetic events that occurred in response to large-scale fluid migration through the regional Knox aquifer. Cathodoluminescence-zoned cements thus provide a framework in which to constrain the origin of the MVT deposits.

Applications to petroleum geology

Accumulations of petroleum and natural gas occur in both siliciclastic sedimentary rocks (mainly sandstones) and carbonate rocks (both limestones and dolomites). To be of economic interest, such accumulations must be contained within a sizeable structural or stratigraphic trap (e.g., an anticline) and occur within reservoir rocks that have sufficient porosity (open pore space) to hold commercially significant quantities of petroleum or natural gas (hydrocarbons). Further, the reservoir rocks must be permeable enough to allow petroleum or natural gas to flow readily into a well bore for production.

Petroleum geologists are particularly interested in the processes and factors that control the creation and destruction of porosity in reservoir rocks because porosity governs the amount of petroleum or natural gas that can be contained within a trap of given size. They are also concerned about the timing of porosity development and the timing of hydrocarbon migration into a trap. Permeability (the ability of a reservoir rock to transmit a fluid) of reservoir rocks also has great significance. Hydrocarbons cannot be produced from reservoir rocks that are impermeable. Fractures in reservoir rocks are particularly important because they greatly increase permeability and fluid flow.

Cathodoluminescence microscopy has several applications to petroleum geology: (1) evaluating porosity reduction owing to compaction and cementation; (2) estimating the timing of cementation; (3) identifying the presence of fractures (which increase permeability) and evaluating the timing of fracture development; (4) mapping diagenetic aureoles above petroleum accumulations, which result from upward seepage of petroleum from traps; and (5) timing of petroleum migration into traps.

Cathodoluminescence-assisted techniques for estimating porosity loss owing to mechanical compaction are described in detail by Houseknecht (1987, 1991), as discussed in Chapter 5. These techniques allow both qualitative and quantitative estimation of compaction in quartz-rich sandstones. Cementation also plays a major role in porosity reduction in both sandstones and carbonate rocks. Cathodoluminescence is particularly useful in identifying and quantifying silica cements in sandstones, which occur mainly as overgrowths on quartz grains. For example, Evans *et al.* (1994) describe a technique for quantification of quartz cement by using combined SEM, CL, and image analysis. Whereas compaction and cementation decrease porosity, fracturing generates porosity and causes a significant increase in permeability of reservoir rocks. The timing of fracture development and any subsequent sealing by cementation relative to the timing of petroleum migration into a trap are exceedingly important. In order for an oil accumulation to take place, potential reservoir rocks must have adequate porosity to hold petroleum (i.e., porosity has not been destroyed by compaction and/or cementation) and be sufficiently permeable to transmit fluids at the time a phase of petroleum migration moves through a petroleum-generating basin.

McLimans (1991) describes a case study of a well core that illustrates use of CL to evaluate the reservoir potential of the Ellenburger Dolomite (Ordovician) in the Val Verde Basin, Texas. This region has undergone several episodes of tectonism, which resulted in extensive fracturing of the Ellenburger Dolomite. The timing of fracture development and any subsequent sealing by cementation relative to the time of hydrocarbon migration are essential to assessing the potential for commercial hydrocarbon accumulation. According to McLimans, CL imaging reveals several generations of fracture development and sealing, not visible by using plane light microscopy. Fractures are cemented by dolomite (red CL) and several generations of calcite cement (dark-brown, light-brown, and yellow CL). Fluid-inclusion data together with CL data indicate four fracture episodes each followed by precipitation of sealing cements in the sequence: dolomite, light-brown (CL) calcite, dark-brown (CL) calcite, and yellow (CL) calcite. The fluid inclusion and CL data indicate that fluids of different composition invaded the fracture porosity of the Ellenburger at different geological times. Unfortunately, study of the cements by photoluminescence techniques did not reveal any evidence of hydrocarbons or oil inclusions. Despite the creation of fracture porosity and the presence of a structural trap, none of the invading fluids that sealed fractures show any evidence for oil migration – presumably indicating that the timing of oil migration did not coincide with the timing of fracture development. Thus, in this case, CL study showed that the Ellenburger Dolomite in this part of the Val Verde Basin has low potential for hydrocarbon accumulation.

Numerous other researchers have reported application of CL microscopy to evaluation of petroleum-reservoir characteristics, particularly to study of porosity and fracture permeability. Barker *et al.* (1991) describe a rather different application of CL microscopy to petroleum geology. These authors used CL microscopy to map regionally zoned carbonate cements in diagenetic aureoles above oil reservoirs in sandstones in the Velma oil field, Oklahoma. These aureoles are areas above petroleum reservoirs where the trace-element concentration of late-diagenetic cements has been altered by upward microseepage of petroleum, which causes strongly reducing conditions to develop above the reservoirs. Barker *et al.* determined the intensity of CL emissions in dolomite and calcite cements and classified the cements as nonluminescing, dull (CL), and bright (CL). As discussed in Chapter 6, CL intensity in carbonate minerals is related to the concentrations of Mn^{2+} activators and Fe^{2+} quencher ions; partitioning of these divalent ions into carbonate cements is favored by reducing conditions. Barker *et al.* observed that carbonate cements in sandstones over the production area are nonluminescent or dull (low Mn^{2+}) and dull to bright (higher Mn^{2+}) on the flanks. Thus, CL emission intensity provides a means for mapping diagenetic alteration above petroleum reservoirs owing to hydrocarbon leakage and consequent inducement of strongly reducing pore-water conditions.

Applications to archeology

Application of CL to archaeological problems is somewhat outside the scope of this book; however, some uses of CL in archeological studies are similar to those in sedimentology and thus may be of interest to sedimentologists. An important area of research in archeology is *sourcing*, that is, determining the geological sources of raw materials used to make the prehistoric artifacts and pottery found at archaeological sites. Cathodoluminescence has been applied to several aspects of archaeological sourcing (provenance) such as tracing marble building stones and statues to their source(s), provenancing of Neolithic stone tools, and characterization of mineral grains in pottery and tracing of these grains to their source(s). For example, Schvoerer *et al.* (1986) suggest that differences in CL color of minerals such as quartz and feldspar may be tied to differences in geological or archeological origin. They indicate that CL can be applied to research on a variety of archeological materials, including ceramics, minerals used for sculpture and architecture, lithic tools, metallic ores, gems, etc., paralleling the use of traditional methods of X-ray diffraction and polarizing microscopy.

To illustrate, Barbin *et al.* (1992) and Barbin (1995) provide a detailed overview of the CL properties of white marbles. They were able to recognize 21 cathodomicrofacies of marbles in classical quarrying areas in Italy, Turkey, Greece, Spain, and France. Each microfacies generally characterizes a given area. These cathodoluminescence facies were grouped into three families on the basis of CL color: (1) orange-luminescence family,

(2) blue-luminescence family, and (3) red-luminescence family. Barbin (1995) states that provenance identification provides information required to reassemble separated parts of artifacts to detect forgeries, to date sculptures, or assess exchanges between populations. See also Lapuente *et al.* (2000).

As an example of CL application to study of lithic artifacts, Akridge and Benoit (2001) describe the CL characteristics of more than 20 chert samples from the Ozark Mountains of the central United States. Examination of chips or flakes from chert artifacts allowed categorization of the cherts into four types: (1) nonluminescent; (2) mainly non-luminescent but with a few luminescent inclusions or fossils; (3) samples with various amounts of orange CL distributed throughout the sample; and (4) samples that exhibit mainly orange CL, which is apparently due to the presence of carbonate minerals in the chert. The authors conclude that CL properties provide a useful criterion for establishing the number of chert varieties and for sourcing chert artifacts collected from archeo-logical sites to their provenance localities.

Julig *et al.* (1998) applied CL to study of the sources of stone artifacts manufactured by Paleo-Indian cultures in the Great Lakes region of North America (*c.* 12 000–7500 years BP). Although fine-grained cryp-tocrystalline chert was used to manufacture some tools, others were made from coarser grained materials, including quartzites and quartz arenites. Geochemical and petrographic techniques are very useful for identifying the source of these coarse materials; however, these techniques may involve partial destruction of the sample. Cathodoluminescence provides an alternative, nondestructive technique for tracing these materials to their sources. In particular, Julig *et al.* determined that samples from Hixton Silicified Sandstone (Cambrian) from bedrock quarry sources could be clearly differentiated from older (Precambrian) sandstones because the detrital quartz grains (dull-red luminescent quartz with 20–30% blue luminescent quartz), early chalcedonic rim cements (non-luminescent), and late pore-fill and rim cement (dull-red luminescence) each have distinct luminescence characteristics. By contrast, samples of the Late Paleoproterozoic Ajibik quartzite exhibit a uniform, weak, dull-red luminescence, which may have developed during metamorphism. Further, they determined that archeological samples from the Cummins and Renier sites are confirmed as Hixton Silicified Sandstone.

Perhaps the most interesting potential archeological application of CL from a sedimentological point of view is the use of CL to trace so-called *temper* in pottery. Temper is sand grains, crushed rock, or ground-up potshards that are added to clay to reduce shrinkage and cracking during drying. Sand grains can be effectively characterized by CL, which may allow the grains to be traced to their geologic sources. Picouet *et al.* (1999) provide an example of the use of cathodoluminescence spectrometry of quartz grains as a tool for ceramic provenance. These authors used SEM–CL spectroscopy to examine quartz grains in pottery from two Neolithic archeological sites, one in Switzerland and one in France. All of the CL spectra of quartz exhibited two components, a blue component centered at about 2.75 eV and a red-orange component centered at about 1.93 eV. By examining the relative intensities of red and blue peaks, the authors were able to identify three types of quartz CL: red, purple, and blue. They were not able to differentiate effectively pottery from the two sites on the basis of quartz CL alone; however, by also characterizing the CL of matrix calcite they could differentiate most of the samples from both sites.

In a different approach to the same problem, Gordon Goles (a colleague from the University of Oregon) initiated a study of quartz grains in potshards from Fiji by using the same method for provenance analysis as that described by Seyedolali *et al.* (1997a). This method is based upon study of CL fabrics in quartz rather than CL color (see discussion in Chapter 4). Basically, volcanic quartz primarily displays zoned CL patterns, plutonic quartz is characterized especially by spider-like fabrics (dark streaks and patches), and metamorphic quartz displays homogeneous or mottled CL. Goles reported (personal communication) that he had made progress in his investigation; preliminary results, on the basis of CL fabric study, indicated that the temper grains in Fiji pottery could be linked to nearby sand dunes. Unfortunately, Gordon died before his work was finished and the results were never published. Nonetheless, this technique for studying temper grains clearly has potential. A possible problem with the technique, as well as the use of CL color in ceramic provenance, has to do with the temperature used in firing pottery. Is it high enough to affect CL emissions in quartz? Cathodoluminescence experiments on fired German quartz sands apparently show that there are no CL changes with temperature (Picouet, 1997; reported in Picouet *et al.*, 1999); however, this potential problem may need further research.

Applications to Precambrian rocks

Most applications of CL to study of sedimentary rocks that we have described in this book are applications to Phanerozoic-age rocks;

however, CL imaging can also be applied to much older, Precambrian rocks. Our own CL work with Precambrian rocks has focused on identifying CL fabrics in metamorphic quartz that can be used in provenance evaluation. For example, Figure 4.8 shows the CL characteristics of Precambrian metamorphic quartz from the Snow Peak area, Idaho. Figure 4.10 shows sheared Precambrian quartz from metavolcanic rocks of the Prescott area, Arizona, and Figure 4.12 depicts spider-like fabrics in Precambrian gneiss adjacent to the Skaergaard Intrusion, Greenland. These examples demonstrate that very old rocks can display CL. As far as we have been able to determine from study of quartz grains in unmetamorphosed rocks, CL fabrics in quartz such as zoning and healed fractures can persist indefinitely unless quartz-bearing rocks are subjected to an episode of greenschist or higher-grade metamorphism.

Other researchers have described various applications of CL to study of Precambrian sedimentary rocks. One common use of CL imaging has been as an adjunct to U–Pb dating of zircons in provenance studies. Cathodoluminescence allows identification of different types of zircon domains that may then be dated *in situ* (see discussion in Chapter 4). For example, Cawood *et al.* (1999) describe Precambrian-age detrital zircons in sedimentary rocks from Mesozoic arc-trench terranes of New Zealand that have U–Pb ages ranging from 1000–1200 Ma. These zircons may have been derived from similar-age igneous bodies along the Median tectonic zone that separates the eastern and western terranes in New Zealand. Cathodoluminescence imaging of these old zircons reveals metamorphic overprints and inherited cores that indicate a complex history.

Ross *et al.* (2001) describe a similar study of zircons from the Precambrian Muskwa assemblage of British Columbia. They report U–Pb ages for detrital zircons, from four sandstone units, that range in age from 3075–1766 Ma, with a pronounced concentration of ages at about 1850 Ma. Cathodoluminescence imaging of the zircons, used to select zones for U–Pb dating, thus shows that CL zoning of zircons can persist for billions of years. The authors conclude that the provenance characteristics (CL and U–Pb ages) of the Muskwa are compatible with sediment derived by unroofing and erosion of the orogenic topography formed by the collisional shortening of the Hearne Province during terminal assembly of the Canadian Shield about 1.8 billion years ago.

A couple of additional case studies serve to illustrate other applications of CL microscopy to Precambrian sedimentary rocks. Ghazban *et al.* (1992) used CL as an accessory technique in their study of multistage dolomitization of the Precambrian Society Cliffs Formation, northern Baffin Island, Northwest Territories, Canada. The Society Cliffs Formation is a largely dolomite unit, 265–850 m thick, which forms part of a 6100 m succession of sediments deposited during the Late Precambrian. Ghazban *et al.* reported that two major stages of dolomitization have occurred: (1) massive dolomitization of precursor carbonates, and (2) late-stage cementation. Dolomites formed during stage 1 have a tan-orange luminescence that is relatively homogeneous. Stage 1 dolomite is overgrown by bright yellow, isopachous-zoned luminescent cements that appear to have formed both during and after brecciation (stage 2). These cements are followed successively by dark red-brown, dull-to bright-yellow nonzoned, bright-red isopachous-zoned, and dull-red luminescent cements. These authors suggest that the fluid involved in stage 1 pervasive dolomitization was seawater modified by mixing with meteoric water in a mixing zone. During the second stage of dolomitization, fractures were filled with a new generation of dolomite precipitated from ascending warm fluids with different chemical compositions, possibly related to those fluids responsible for deposition of associated Pb–Zn deposits.

Fairchild *et al.* (1991) used CL microscopy to study early cements in coastal deposits of the Late Precambrian Draken Formation, Spitsbergen. The deposits included quartz sandstones, stromatolitic mats, conglomerates with silicified intraclasts, dolomite conglomerates with desiccated mudrocks, oolitic/pisolitic grainstones and fenstral dolomites. On the basis of CL characteristics, the authors recognized six types of cements: (1) rare mosaics of equant calcite microspar, which displays consistent CL zonation – interpreted as neomorphosed submarine cement (originally aragonite); (2) rare isopachous fibrous dolomite that displays dark CL with some fuzzy parallel bands – interpreted as submarine cement (originally Mg-calcite); (3) rare dolomicrite filling interstices in oolites, which displays uniform or patchy CL – interpreted as submarine hardground (originally Mg-calcite); (4) dolomicrite with irregular distribution concentrating at grain contacts, which displays uniform CL – interpreted as vadose (meniscus) cement (originally high magnesian calcite); (5) displacive dolomicrite and dolomicrospar, which display bright or zoned CL – interpreted as emergent horizons (calcrete) deposited by highly supersaturated vadose solutions; and (6) complex crusts up to millimeter-scale composed of mammilated dolomicrite and fibrous dolomite, commonly displaying CL zonation – interpreted as emergent horizons subjected to artesian diagenesis by highly supersaturated solutions, which were microbially influenced. These diagenetic fabrics indicate that dolomitization was a significant syndepositional process that altered original metastable carbonates. The CL data, taken together with petrographic and biogenic data, suggest that the Draken Formation was deposited in a spectrum of peritidal environments including ooid shoals, protected subtidal zones, tidal sandflats and protected carbonate mudflats.

These case histories indicate that CL microscopy may be useful in a variety of applications to Precambrian sedimentary rocks, ranging from provenance analysis to evaluation of diagenetic features, in much the same way that it is applied to younger sedimentary rocks. As mentioned, available evidence appears to confirm that CL fabrics such as growth zoning in zircon can persist for billions of years unless homogenized by thermal metamorphism. We are not aware of any research that has evaluated the effects of geological age on CL color. It seems likely that the wavelengths of CL emissions also persist unchanged in the absence of thermal metamorphism. Researchers who use CL color of quartz as a provenance tool apparently make that assumption; however, hard evidence to support the assumption is lacking.

References

Adams, A. E. and W. S. MacKenzie, 1998. A Color Atlas of Carbonate Sediments and Rocks Under the Microscope, New York, NY, John Wiley & Sons, Inc.

Akridge, D. G. and P. H. Benoit, 2001. Luminescence properties of chert and some archaeological applications. Journal of Archaeological Science, 28, 143–51.

Anderson, A. T., Jr., A. M. Davis, and F. Lu, 2000. Evolution of Bishop Tuff rhyolitic magma based on melt and magnetite inclusions, and zoned phenocrysts. Journal of Petrology, 41, 449–73.

Arlinghaus, H. F., 2002. Static secondary ion mass spectrometry (SSIMS). In Bubert, H. and H. Jenett (eds.), Surface and Thin Film Analysis, Weinheim, Wiley-VCH Verlag GmbH, pp. 86–106.

Barbarand, J. and M. Pagel, 2001. Cathodoluminescence study of apatite crystals. American Mineralogist, 86, 473–84.

Barbin, V., 1995. Cathodoluminescence of carbonates: new applications in geology and archaeology. In Re?dmond, G. L. Balk, and D. J. Marshall (eds.), Luminescence: Scanning Microscopy Supplement 9, Chicago, Scanning Microscopy International, pp. 113–23.

Barbin, V., 2000. Cathodoluminescence of carbonate shells: biochemical vs. diagenetic process. In Pagel, M., V. Barbin, P. Blanc, and D. Ohnenstetter (eds.), Cathodoluminescence in Geosciences, Berlin, Springer-Verlag, pp. 303–29.

Barbin, V., K. Ramseyer, D. Decrouez, et al., 1992. Cathodoluminescence of white marbles: an overview. Archaeometry, 34, 175–83.

Barker, C. E. and T. Wood, 1986. A review of the Technosyn and Nuclide cathodoluminescence stages and their application to sedimentary geology. In Hagni R. D. (ed), Process Mineralogy, VI, Warrendale, PA, The Metallurgical Society, Inc., pp. 137–158.

Barker, C. E., D. K. Higley, and M. C. Dalziel, 1991. Using cathodoluminescence to map regionally zoned carbonate cements occurring in diagenetic aureoles above oil reservoir: initial results from the Velma oil field, Oklahoma. In Barker, C. E. and O. C. Kopp (eds.), Luminescence Microscopy and Spectroscopy: Qualitative and Quantitative Applications, SEPM Short Course 25, pp. 155–60.

Behr, H. J., 1989. Die geologische Aktivita?t von Krustenfluiden. In *Gesteinsfluide – Ihre Herkunft und Bedeutung für Geologische Prozesse*, Hannover, Niedersa?chsische Akademie der Geowissenschaffer, pp. 7–42.

Benninghoven, A., F. G. Rüdenauer, and H. W. Werner, 1987. Secondary Ion Mass Spectrometry: Basic Concepts, Instrumental Aspects, Applications and Trends, New York, NY, John Wiley & Sons.

Bernet, M. and K. Bassett, 2005. Provenance analysis by single-quartz grain SEM–CL/optical microscopy. Journal of Sedimentary Research, 75, 492–500.

Best, M. G. and E. H. Christiansen, 1997. Origin of broken phenocrysts in ash-flow tuffs. Geological Society of America Bulletin, 109, 63–73.

Boggs, S., Jr., 1992. Petrology of Sedimentary Rocks, New York, Macmillan Publishing Co.

Boggs, S., Jr., 2001. Principles of Sedimentology and Stratigraphy, 3rd edn., Upper Saddle River, NJ, Prentice Hall.

Boggs, S., Jr., D. H. Krinsley, G. G. Goles, A. Seyedolali, and H. Dypvik, 2001. Identification of shocked quartz by scanning cathodoluminescence imaging. Meteoritics & Planetary Sciences, 36, 783–91.

Boggs, S., Jr., Y.-I. Kwon, G. G. Goles, et al., 2002. Is quartz cathodoluminescence color a reliable provenance tool? A quantitative examination. Journal of Sedimentary Research, 72, 408–15.

Bourque, P.-A., M. M. Savard, G. Chi, and P. Dansereau, 2001. Diagenesis and porosity evolution of the Upper Silurian-lowermost Devonian West Point reef limestone, eastern Gaspe? Belt, Que?bec Appalachians. Bulletin of Canadian Petroleum Geology, 49, 299–326.

Breton, P. J., 1999. From microns to nanometres: early landmarks in the science of scanning electron microscope imaging. Scanning Microscopy, 13, 1–6.

Bruckschen, P., R. D. Neuser, and D. K. Richter, 1992. Cement stratigraphy in Triassic and Jurassic limestones of the Weserbergland (northwestern Germany). Sedimentary Geology, 81, 195–214.

Budd, D. A., U. Hammes, and W. B. Ward, 2000, Cathodoluminescence in calcite cements: new insights on Pb and Zn sensitizing, Mn activation, and Fe quenching at low trace-element concentrations. Journal of Sedimentary Research, 70, 217–26.

Burley, S. D. and R. H. Worden, 2003. Sandstone Diagenesis: Recent and Ancient, Reprint Series Volume 4 of the International Association of Sedimentologists, Oxford, Blackwell Publishing Ltd.

Campbell, J. L. and G. K. Czamanske, 1998. Micro-PIXE in earth science. In Applications of Microanalytical Techniques to Understanding Mineralizing Processes, Reviews in Economic Geology, 7, pp. 169–85.

Carlson, R. C., R. H. Goldstein, and P. Enos, 2003. Effects of subaerial exposure on porosity evolution in the Carboniferous Lisburne Group, northeastern Brooks Range, Alaska, USA. In Permo-Carboniferous Carbonate Platforms and Reefs, SEPM Special Publication 78 and AAPG Memoir 83, pp. 269–90.

Cawood, P. A., A. A. Nemchin, A. Leverenz, A. Saeed. and P. F. Ballance, 1999. U/Pb dating of detrital zircons: implications for the provenance record of

Gondwana margin terranes. Geological Society of America Bulletin, 111, 1107–19.

Coniglio, M., 1989. Neomorphism and cementation in ancient deep-water limestones, Cow Head Group (Cambro-Ordovician), western Newfoundland, Canada. Sedimentary Geology, 65, 15–33.

Dapples, E. C., 1979, Diagenesis in sandstones. In Larsen, G., and G. V. Chilingar (eds.), Diagenesis in Sediments and Sedimentary Rocks, Developments in Sedimentology 25A, Amsterdam, Elsevier Scientific Publishing Company, pp. 31–97.

Dickinson, W. W. and K. L. Milliken, 1995. The diagenetic role of brittle deformation in compaction and pressure solution, Etjo Sandstone, Nambia. Journal of Geology, 103, 339–47.

D'Lemos, R. S., A. T. Kearsley, J. W. Pemboke, G. R. Watt, and P. Wright, 1997. Complex quartz growth histories in granite revealed by scanning cathodoluminescence techniques. Geological Magazine, 134, 549–52.

Dorobek, S. L., 1987. Petrography, geochemistry, and origin of burial diagenetic facies, Siluro-Devonian Helderberg Group (carbonate rocks), central Appalachians. American Association of Petroleum Geologists Bulletin, 71, 492–514.

Dromgoole, E. L. and Walter, L. M., 1990. Iron and manganese incorporation into calcite: effects of growth kinetics, temperature and solution chemistry. Chemical Geology, 81, p. 311–36.

Durocher, S. and I. S. Al-Aasm, 1997. Dolomitization and neomorphism of Mississippian (Visean) Upper Debolt Formation, Blueberry Field, northeastern British Columbia: geologic, petrologic, and chemical evidence. American Association of Petroleum Geologists Bulletin, 81, 954–77.

Ebers, M. L. and O. C. Kopp, 1979. Cathodoluminescent microstratigraphy in gangue dolomite, the Mascot-Jefferson City District, Tennessee. Economic Geology, 74, 908–18.

El Ali, A., V. Barbin, G. Calas, B. Cervelle, K. Ramseyer, and J. Bouroulec, 1993. Mn^{2+}-activated luminescence in dolomite, calcite and magnesite: quantitative determination of manganese and site distribution by EPR and CL spectroscopy. Chemical Geology, 104, 189–202.

Emery, D. and J. D. Marshall, 1989. Zone calcite cements: has analysis outpaced interpretation? Sedimentary Geology, 65, 205–10.

Evamy, B. D., 1969. The precipitational environment and correlation of some calcite cements deduced from artificial staining. Journal of Sedimentary Petrology, 39, 787–821.

Evans, J., A. J. C. Hogg, M. S. Hopkins, and R. J. Howarth, 1994. Quantification of quartz cements by using combined SEM, CL, and image analysis. Journal of Sedimentary Research, A64, 334–8.

Fairchild, I. J., A. H. Knoll, and K. Swett, 1991. Coastal lithofacies and biofacies associated with syndepositional dolomitization and silicification (Draken Formation, Upper Riphean, Svalbard). Precambrian Research, 53, 165–97.

Finch, A. A. and Klein, J., 1999. The causes and petrological significance of cathodoluminescence emissions from alkali feldspars. Contributions to Mineralogy and Petrology, 135, 234–43.

Flem, B., R. B. Larsen, A. Grimstvedt, and J. Mansfeld, 2002. *In situ* analysis of trace elements in quartz by using laser ablation inductively coupled plasma mass spectrometry. Chemical Geology, 182, 237–47.

Folk, R. L., 1965. Some aspects of recrystallization in ancient limestones. In Dolomitization and Limestone Diagenesis, Society of Economic Paleontologists and Mineralogists Special Publication, 13, pp. 14–48.

Fournier, R. O., 1999. Hydrothermal processes related to movement of fluid from plastic to brittle rock in the magmatic–epithermal environment. Economic Geology, 94, 1193–212.

Fraser, D. G., 1995. The nuclear microprobe – PIXE, PIGE, RBS, NRA and ERDA. In Potts, P. J., J. F. W. Bowles, S. J. B. Reed, and M. R. Cave (eds.), Microprobe Techniques in the Earth Sciences, London, Chapman & Hall, pp. 140–62.

Gaft, M., R. Reisfeld, G. Panczer, *et al.* 1997. Accommodation of rare-earth and manganese by apatite. Optical Materials, 8, 149–56.

Ghazban, F., H. P. Schwarcz, and D. C. Ford, 1992. Multistage dolomitization of the Society Cliffs Formation, northern Baffin Island, Northwest Territories, Canada. Canadian Journal of Earth Science, 29, 1459–73.

Gilhaus, A. and D. K. Richter, 2001. Polyphase Dolomitgenese in oberpermischen und obertrissischen Sabkha-Kleinzyklen von Hydra (Griechenland). Neues Jahrbuch für Geologie und Palaeontologie, Monatshefte, 2001, 399–422.

Goldstein, J. I., D. E. Newbury, P. Echlin *et al.*, 2003. Scanning Electron Microscopy and X-Ray Microanalysis, 3rd edn., New York, NY, Kluwer Academic/Plenum Publishers.

Goldstein, R. H., 1988. Cement stratigraphy of Pennsylvanian Holder Formation, Sacramento Mountains, New Mexico. American Association of Petroleum Geologists Bulletin, 72, 425–38.

Goldstein, R. H., 1991. Practical aspects of cement stratigraphy with illustrations from Pennsylvanian limestone and sandstone, New Mexico and Kansas. In *Luminescence Microscopy and Spectroscopy: Qualitative and Quantitative Applications*, SEPM Short Course 25, pp. 123–131.

Goldstein, R. H. and C. Rossi, 2002. Recrystallization in quartz overgrowths. Journal of Sedimentary Research, 72, 432–40.

Gorobets, B. S. and G. Walker, 1995. Origins and luminescence in minerals: A summary of fundamental studies and applications. In Marfunmin, A. S. (ed.), Advanced Mineralogy 2, Methods and Instrumentations: Results and Recent Developments, Berlin, Springer-Verlag, pp. 138–46.

Götte, Th., R. D. Neuser, and D. K. Richter, 2001. New parameters of quartz in sandstone-petrography: cathodoluminescence (CL)-investigation of mature sands and sandstones of north-western Germany. Abstracts of CL 2001 in Freiberg/Sachen, Germany, pp. 38–9.

Götze, J., 2002. Potential of cathodoluminescence (CL) microscopy and spectroscopy for the analysis of minerals and materials. Analytical and Bioanalytical Chemistry, 374, 703–8.

Götze, J. and M. Magnus, 1997. Quantitative determination of mineral abundance in geological samples using combined cathodoluminescence microscopy and image analysis. European Journal of Mineralogy, 9, 1207–15.

Götze, J. and W. Zimmerle, 2000. Quartz and silica as guide to provenance in sediments and sedimentary rocks. Contributions to Sedimentary Geology, 21, pp. 1–91.

Götze, J., M. R. Krbetscek, D. Habermann, and D. Wolf, 2000. High-resolution cathodoluminescence studies of feldspar minerals. In Pagel, M., V. Barbin, P. Blanc, and D. Ohnenstetter (eds.), Cathodoluminescence in Geosciences, Berlin, Springer-Verlag, pp. 245–70.

Götze, J., M. Plötze, and D. Habermann, 2001. Origin, spectral characteristics and practical applications of the cathodoluminescence (CL) of quartz – a review. Mineralogy and Petrology, 71, 225–50.

Götze, J., M. Plötze, Th. Götte, R. D. Neuser, and D. K. Richter, 2002. Cathodoluminescence (CL) and electron paramagnetic resonance (EPR) studies of clay minerals. Mineralogy and Petrology, 76, 195–212.

Graton, L. C. and Fraser, 1935. Systematic packing of spheres with particular relation to porosity and permeability. Journal of Geology, 43, 785–909.

Gratz, A., D. K. Fisler, and B. F. Bohor, 1996. Distinguishing shocked from tectonically deformed quartz by use of the SEM and chemical etching. Earth and Planetary Science Letters, 142, 513–21.

Grover, G., Jr. and J. F. Read, 1983. Paleoaquifer and deep burial related cements defined by regional cathodoluminescence patterns, Middle Ordovician carbonates, Virginia. American Association of Petroleum Geologists Bulletin, 67, 1275–303.

Habermann, D., R. D Neuser, and D. R. Richter, 1996. REE-activated cathodoluminescence of calcite and dolomite: high-resolution spectrometric analysis of CL-emission (HRS–CL). Sedimentary Geology, 101, 1–7.

Habermann, D., R. D. Neuser, and D. K. Richter, 1998. Lower limit of Mn^{2+}-activated cathodoluminescence of calcite: state of the art. Sedimentary Geology, 116, 13–24.

Habermann, D., R. D. Neuser, and D. K. Richter, 2000a. Quantitative high resolution analysis of Mn^{2+} in sedimentary calcite. In Pagel, M., V. Barbin, P. Blanc, and D. Ohnenstetter (eds.), Cathodoluminescence in Geosciences, Berlin, Springer Verlag, pp. 331–58.

Habermann, D., T. Götte, J. Meijer et al., 2000b. High resolution rare-earth elements analyses of natural apatite and its application in geo-sciences: combined micro-PIXE, quantitative CL spectroscopy and electron spin resonance analyses. Nuclear Instruments and Methods in Physics Research B, 161–163, 846–51.

Hagni, R. D., 1986. Importance of cathodoluminescence microscopy in study of sedimentary ironstones. American Association of Petroleum Geologists Bulletin, 70, 598.

Hartmann, B. H., K. Juha?sz-Bodna?r, K. Ramseyer, and A. Matter, 2000. Polyphased quartz cementation and its sources: a case study from the Upper Palaeozoic Haushi Group sandstones, Sultanate of Oman. In Quartz Cementation in Sandstones, International Association of Sedimentologists, Special Publication, 29, pp. 253–70.

Heaney, P. J., 1994. Structure and chemistry of the low-pressure polymorphs. In Silica: Physical Behavior, Geochemistry and Material Applications, Mineralogical Society of America Reviews in Mineralogy, 29, pp. 1–40.

Herzog, L. F., D. J. Marshall, and R. R. Babione, 1970. The Luminoscope – a new instrument for studying the electron-stimulated luminescence of terrestrial, extra-terrestrial and synthetic materials under the microscope. In Weber, J. N. and E. White (eds.), Space Science Applications of Solid State Luminescence Phenomena, Materials Research Laboratory Special Publication, 70–101, pp. 79–98.

Hinton, R. W., 1995. Ion microprobe analysis in geology. In Potts, P. J., J. F. W. Bowles, S. J. B. Reed, and M. R. Cave (eds.), Microprobe Techniques in the Earth Sciences: London, Chapman & Hall, pp. 237–89.

Hogg, A. J. C., E. Sellier, and A. J. Jourdan, 1992. Cathodoluminescence of quartz cements in Brent Group sandstones, Alwyn South, UK North Sea. In Geology of the Brent Group, Geological Society Special Publication No. 61, pp. 421–40.

Hoholick, J. D., 1984. Regional variations of porosity and cement: St. Peter and Mount Simon sandstones in Illinois Basin. American Association of Petroleum Geologists Bulletin, 68, 753–64.

Houseknecht, D. W., 1987. Assessing the relative importance of compaction processes and cementation to reduction of porosity in sandstones. American Association of Petroleum Geologists Bulletin, 71, 633–42.

Houseknecht, D. W., 1991. Use of cathodoluminescence petrography for understanding compaction, quartz cementation, and porosity in sandstones. In Luminescence Microscopy and Spectroscopy: Quantitative and Qualitative Applications, SEPM Short Course 25, pp. 59–66.

Hutter, H., 2002. Dynamic secondary ion mass spectrometry. In Bubert, H. and H. Jenett (eds.), Surface and Thin Film Analysis, Weinheim, Wiley-VCH Verlag GmbH, pp. 106–21.

Jackson, S. E., H. P. Longerich, G. R. Dunning, and B. J. Fryer, 1992. The application of laser-ablation microprobe–inductively coupled plasma–mass spectrometry (LAM–ICP-MS) to *in situ* trace-element determinations in minerals. Canadian Mineralogist, 30, 1049–64.

Jarvis, K. E., A. L. Gray, and R. S. Houk, 1992. Handbook of Inductively Coupled Plasma Mass Spectrometry, Glasgow, Blackie & Sons Ltd.

Johnson, S. A. E., J. L. Campbell, K. G. Malmqvist (eds.), 1995. Particle-Induced X-Ray Emission Spectrometry (PIXE), New York, NY, Wiley Interscience.

Julig, P. J., D. G. F. Long, and R. G. V. Hancock, 1998. Cathodoluminescence and petrographic techniques for positive identification of quartz-rich lithic artifacts from late Paleo-Indian sites in the Great Lakes region. The Wisconsin Archeologist, 79, 68–88.

Kaufman, J., H. S. Cander, L. D. Daniels, and W. J. Meyers, 1988. Calcite cement stratigraphy and cementation history of the Burlington–Keokuk Formation (Mississippian), Illinois and Missouri. Journal of Sedimentary Petrology, 58, 312–26.

Keller, T. J., J. M. Gregg, and K. L. Shelton, 2000. Fluid migration and associated diagenesis in the greater Reelfoot Rift region, Midcontinent, United States. Geological Society of America Bulletin, 112, 1680–93.

Kempe, U. and J. Götze, 2002. Cathodoluminescence (CL) behavior and crystal chemistry of apatite from rare-metal deposits. Mineralogical Magazine, 66, 151–72.

Kempe, U., T. Grunder, L. Nasdala, and D. Wolf, 2000. Relevance of cathodoluminescence for the interpretation of U–Pb zircon ages, with an example of an application to a study of zircons from the Saxonian Granulite Complex, Germany. In Pagel, M., V. Barbin, P. Blanc, and D. Ohnenstetter (eds.), Cathodoluminescence in Geosciences, Berlin, Springer-Verlag, pp. 415–55.

Kopp O. C., 1991. Studies of ore deposits and trace elements in minerals. In Luminescence Microscopy and Spectroscopy, Qualitative and Quantitative Applications, SEPM Short Course 25, pp. 117–22.

Kopp, O. C., M. L. Ebers, L. B. Cobb, *et al.*, 1986. Application of cathodoluminescence microscopy to the study of gangue carbonates in Mississippi Valley-type deposits in Tennessee: the search for a "Tennessee Trend." In Hagni, R. D. (ed.), Process Mineralogy, VI, Warrendale, PA, The Metallurgical Society Inc., pp. 53–67.

Kopp, O. C., E. L. Fuller, Jr., and M. R. Owen, 1995. Interpretation of cathodoluminescence spectra obtained from dolomite and calcite gangue minerals, and dolostone breccias in the Central Tennessee Zinc District (USA). In *Luminescence*, Scanning Microscopy Supplement 9, pp. 211–23.

Krinsley, D. H. and P. W. Hyde, 1971. Cathodoluminescence studies of sediments. Scanning Electron Microscopy/1971 Part I, Proceedings of the Fourth Annual Scanning Electron Microscopy Symposium, Chicago, IL, IIT Research Institute, pp. 409–16.

Krinsley, D. and N. K. Tovey, 1978. Cathodoluminescence in Quartz Sand Grains. Scanning Electron Microscopy, 1, pp. 887–94.

Krinsley, D. H., K. Pye, S. Boggs, Jr., and N. K. Tovey, 1998, Backscattered Electron Microscopy and Image Analysis of Sediments and Sedimentary Rocks, Cambridge, Cambridge University Press.

Kupecz, J. A. and L. S. Land, 1994. Progressive recrystallization and stabilization of early-stage dolomite: Lower Ordovician Ellenburger Group, west Texas. In Dolomites: A Volume in Honour of Delomieu, International Association of Sedimentologists, Special Publication, No. 21, pp. 255–79.

Kwon, Y-N and S. Boggs, Jr., 2002. Provenance interpretation of Tertiary sandstones from the Cheju Basin (NE East China Sea): a comparison of conventional petrographic and scanning cathodoluminescence techniques. Sedimentary Geology, 152, 29–43.

Lapuente, M. P., B. Turi, and P. Blanc, 2000. Marbles from Roman Hispania: stable isotope and cathodoluminescence characterization. Applied Geochemistry, 15, 1469–93.

Laubach, S. E., 1997. A method to detect natural fracture strike in sandstones. American Association of Petroleum Geologists Bulletin, 81, 604–23.

Lee, M. R., 2000. Imaging of calcite by optical and SEM cathodoluminescence. Microscopy and Analysis, 70, 15–16.

Lee, M. R. and G. M. Harwood, 1989. Dolomite calcitization and cement zonation related to uplift of the Raisby Formation (Zechstein carbonate), northeast England. Sedimentary Geology, 65, 285–305.

Lee, M. R., R. W. Martin, C. Trager-Cowan, and P. R. Edwards, 2005. Imaging of cathodoluminescence zoning in calcite by scanning electron microscopy and hyperspectral mapping. Journal of Sedimentary Research, 75, 313–22.

Lev, S. M., S. M. McLennan, W. J. Meyers, and G. N. Hanson, 1998. A petrographic approach for evaluating trace-element mobility in a black shale. Journal of Sedimentary Research, 68, 970–80.

Leverenz, H. W., 1968. An Introduction to Luminescence of Solids, New York, NY, Dover Publications.

Lohmann, K. C. and J. C. G. Walker, 1989. The $\delta^{18}O$ record of Phanerozoic abiotic marine calcite cements. Geophysical Research Letters, 16, 319–22.

Long, J. V. P., 1963. Recent advances in electron-probe analysis. In Mueller, W. M. and M. Fay (eds.), Advances in X-Ray Analysis: *Proceedings of the 11th Annual Conference on Applications of X-ray Analysis, August 1962*, New York, NY, Plenum Press, vol. 6, pp. 276–90.

Long, J. V. P. and S. O. Agrell, 1965. The cathodo-luminescence of minerals in thin section. Mineralogical Magazine, 34, 318–26.

Lowenstam, H. A. and S. Weiner, 1989. On Biomineralization, Oxford, Oxford University Press.

Lyon, I. C., S. D. Burley, P. J. McKeever, J. M. Saxton, and C. Macaulay, 2000. Oxygen isotope analysis of authigenic quartz in sandstones: a comparison of ion microprobe and conventional analytical techniques. In *Quartz Cementation in Sandstones*, International Association of Sedimentologists Special Publication, 29, pp. 299–316.

Machel, H.G., 1985. Cathodoluminescence in calcite and dolomite and its chemical interpretation. Geoscience Canada, 12, 139–47.

Machel, H.G., 2000. Application of cathodoluminescence to carbonate diagenesis. In Pagel, M., V. Barbin, P. Blanc, and D. Ohnenstetter (eds.); Cathodoluminescence in Geosciences, Berlin, Springer-Verlag, pp. 271–301.

Machel, H. G. and E. A. Burton, 1991. Factors governing cathodoluminescence in calcite and dolomite, and their implications for studies of carbonate diagenesis. In *Luminescence Microscopy and Spectroscopy: Qualitative and Quantitative Applications*, SEPM Short Course 25, pp. 37–57.

Machel, H.G., R. A. Mason, A. N. Mariano, and A. Mucci, 1991. Causes and emission of luminescence in calcite and dolomite. In *Luminescence Microscopy and Spectroscopy: Qualitative and Quantitative Applications*, SEPM Short Course 25, pp. 9–25.

MacRae, N. D., 1995. Secondary-ion mass spectrometry and geology. The Canadian Mineralogist, 33, 219–36.

Major, R.P., 1991. Cathodoluminescence in Post-Miocene carbonates, In *Luminescence Microscopy and Spectroscopy: Qualitative and Quantitative Applications*, SEPM Short Course 25, pp. 149–53.

Makowitz, A. and K. L. Milliken, 2003. Quantification of brittle deformation in burial compaction, Frio and Mount Simon Formation sandstones. Journal of Sedimentary Research, 73, 1007–21.

Maliva, R. G., 1989. Displacive syntaxial overgrowths in open marine limestones. Journal of Sedimentary Petrology, 59, 397–403.

Mariano, A. N., 1988. Some further geological applications of cathodoluminescence. In Marshall, D. J., Cathodoluminescence of Geological Materials, Boston, MA, Unwin Hyman, pp. 94–123.

Marshall, D. J., 1988. Cathodoluminescence of Geological Materials, Boston, Unwin Hyman.

Marshall, D. J., 1991. Combined cathodoluminescence and energy dispersive spectroscopy. In *Luminescence Microscopy and Spectroscopy: Qualitative and Quantitative Applications*, SEPM Short Course 25, pp. 27–35.

Marshall, D. J., 1993. The present status of cathodoluminescence attachments for optical microscopes. Scanning Microscopy, 7, 861–74.

Marshall, D. J., J. H. Giles, and A. Marino, 1988. Combined instrumentation for EDS elemental analysis and cathodoluminescence studies of geological materials. In Hagni, R. D. (ed.), Process Mineralogy VI, Warrendale, PA, The Metallurgical Society Inc., pp. 117–35.

Matter, A. and K. Ramseyer, 1985. Cathodoluminescence microscopy as a tool for provenance studies of sandstones. In Zuffa, G. G. (ed.), Provenance of Arenites, Dordrecht, D. Reidel Publishing Co., pp. 191–211.

McLimans, R. K., 1991. Studies of reservoir diagenesis, burial history, and petroleum migration using luminescence micrography. In *Luminescence Microscopy and Spectroscopy: Qualitative and Quantitative Applications*, SEPM Short Course 25, pp. 97–106.

Meyers, W. J., 1991. Calcite cement stratigraphy: an overview. In *Luminescence Microscopy and Spectroscopy: Qualitative and Quantitative Applications*, SEPM Short Course 25, pp. 133–48.

Meyers, W. J., 1974. Carbonate cement stratigraphy of the Lake Valley Formation (Mississippian), Scaramento Mountains, New Mexico. Journal of Sedimentary Petrology, 44, 837–61.

Meyers, W. J., 1978. Carbonate cements: their regional distribution and interpretation in Mississippian limestones of southwestern New Mexico. Sedimentology, 25, 371–400.

Miller, J., 1988. Cathodoluminescence microscopy. In Tucker, M. (ed.), Techniques in Sedimentology, Oxford, Blackwell Scientific Publications, pp. 174–90.

Milliken, K. L., 1994. Cathodoluminescence textures and the origin of quartz silt in Oligocene mudrocks, south Texas. Journal of Sedimentary Research, A64, 567–71.

Milliken, K. L. and S. E. Laubach, 2000. Brittle deformation in sandstone diagenesis revealed by scanned cathodoluminescence imaging with application to characterization of fractured reservoirs. In Pagel, M., V. Barbin, P. Blanc, and D. Ohnenstetter (eds.), Cathodoluminescence in Geosciences, Berlin, Springer Verlag, pp. 225–43.

Mitchell, R. H., J. Xiong, A. N. Mariano, and M. E. Fleet, 1997. Rare-earth-element-activated cathodoluminescence in apatite. The Canadian Mineralogist, 35, 979–98.

Montañez, I. P., 1997. Application of cathodoluminescencent cement stratigraphy for delineating regional diagenetic and fluid migration events associated with Mississippi Valley-type mineralization in the southern Appalachians. Special Publication, Society of Economic Geologists, 4, 432–47.

Muir, M. D. and P. R. Grant, 1974. Cathodoluminescence. In Holt, D. B. and M. D. Muir, Quantitative Scanning Electron Microscopy, London, Academic Press, pp. 287–334.

Müller, A., 2000. Cathodoluminescence and characterisation of defect structures in quartz with applications to the study of granitic rocks. Doctoral dissertation, Georg-August-Universita?t zu Göttingen.

Müller, A., R. Seltmann, and H.-J. Behr, 2000. Application of cathodoluminescence to magmatic quartz in a tin granite – case study from the Schellerhau Granite complex, eastern Erzgebirge, Germany. Mineralium deposita, 35, 169–89.

Müller, A, M. Wiedenbeck, A. M. van den Kerkhof, A. Kronz, and K. Simon, 2003. Trace elements in quartz – a combined electron microprobe, secondary ion mass spectrometry, laser-ablation ICP–MS, and cathodoluminescence study. European Journal of Mineralogy, 15, 747–63.

Neuser, R. D., D. K. Richter, and A. Vollbrecht, 1989. Natural quartz with brown/violet cathodoluminescence – genetic aspects evident from spectral analysis. Zentralblatt für Geologie und Palaeontologie, Teil I: Allgemeine, Angewandte, Regionale und Historische Geologie, 1988 (7–8), 919–30.

Nielsen, P., R. Swennen, and E. Keppens, 1994. Multiple-step recrystallization within massive ancient dolomite units; an example from the Dinantian of Belgium: Sedimentology, 41, 567–84

Oatley, C. W., 1972. The Scanning Electron Microscope, Cambridge, Cambridge University Press.

Oatley, C. W., 1982. The early history of the scanning electron microscope. Journal of Applied Physics, 53, R1–13.

Onasch, C. M. and T. W. Vennemann, 1995. Disequilibrium partitioning of oxygen isotopes associated with sector zoning in quartz. Geology, 23, 1103–6.

Owen, M. R., 1991. Application of cathodoluminescence to sandstone provenance. In Luminescence Microscopy and Spectroscopy: Qualitative and Quantitative Applications, SEPM Short Course 25, pp. 67–75.

Owen, M. R. and A. V. Carozzi, 1986. Southern provenance of upper Jackfork Sandstone, southern Ouachita Mountains: cathodoluminescence petrology. Geological Society of America, Bulletin, 97, 110–15.

Padovani, E. R., S. B. Shirley, and G. Simmons, 1982. Characteristics of microcracks in amphibolite and granulite facies grade rocks from southeastern Pennsylvania. Journal of Geophysical Research, 87, 8605–30.

Pagel, M., V. Barbin, P. Blanc, and D. Ohnenstetter, 2000a. Cathodo-luminescence in geosciences: an introduction. In Pagel, M., V. Barbin,

P. Blanc, and D. Ohnenstetter (eds.), Cathodoluminescence in Geosciences, Berlin, Springer-Verlag, pp. 1–21.

Pagel, M., V. Barbin, P. Blanc, and D. Ohnenstetter (eds.), 2000b. Cathodoluminescence in Geosciences, Berlin, Springer-Verlag.

Passchier, C. W. and R. A. J. Trouw, 1996. Microtectonics, Berlin, Springer-Verlag.

Paxton, S. T., J. O. Szabo, J. M. Ajdukiewicz, and R. E. Klimentidis, 2002. Construction of an intergranular volume compaction curve for evaluating and predicting compaction and porosity loss in rigid-grain sandstone reservoirs. American Association of Petroleum Geologists Bulletin, 86, 2047–67.

Penniston-Dorland, S. C., 2001. Illumination of vein quartz textures in a porphyry copper ore deposit using scanned cathodoluminescence: Grasberg Igneous Complex, Irian Jaya, Indonesia. American Mineralogist, 86, 652–66.

Pennock, G. M., 1995. Scanning electron microscopy and image formation. In Marfunin, A. S. (ed.), Advanced Minerlogy: Methods and Instrumentation. Results and Recent Developments, Berlin, Springer Verlag, vol. 2, pp. 273–9.

Peppard, B. T., I. M. Steele, A. M. Davis, P. J. Wallace, and A. T. Anderson, 2001. Zoned quartz phenocrysts from the rhyolitic Bishop Tuff. American Mineralogist, 86, 1034–52.

Perkins, W. T. and N. J. G. Pearce, 1995. Mineral microanalysis by microprobe inductively coupled plasma mass spectrometry. In Potts, P. J., J. F. W. Bowles, S. J. B. Reed, and M. R. Cave (eds.), Microprobe Techniques in the Earth Sciences, London, Chapman & Hall, pp. 291–325.

Perny, B., P. Eberhardt, K. Ramseyer, J. Mullis, and R. Pankrath, 1992. Microdistribution of Al, Li, and Na in α-quartz: possible causes and correlation with short lived cathodoluminescence. American Mineralogist, 77, 534–44.

Pettijohn, F. J., P. E. Potter, and R. Siever, 1973. Sand and Sandstone, New York, NY, Springer-Verlag.

Picouet, P., 1997. Application de la cathodoluminescence a? l'e?tude des ce?ramiques modernes et arche?ologiques. Ph.D. Thesis, University of Fribourg, Switzerland.

Picouet, P., M. Maggetti, D. Piponnier, and M. Schvoerer, 1999. Cathodoluminescence spectroscopy of quartz grains as a tool for ceramic provenance. Journal of Archaeological Science, 26, 943–9.

Poller, U., 2000. A combination of single zircon dating by TIMS and cathodoluminescence investigations of the same grain: the CLC method – U–Pb geochronology for metamorphic rocks. In Pagel, M., V. Barbin, P. Blanc, and D. Ohnenstetter (eds.), Cathodoluminescence in Geosciences, Berlin, Springer-Verlag, pp. 401–14.

Popp, B. N., T. F. Anderson, and P. A. Sandberg, 1986. Brachiopods as indicators of original isotopic composition in some Paleozoic limestones. Geological Society of America Bulletin, 97, 1262–9.

Ramsey, J. G., 1980. The crack–seal mechanism of rock deformation. Nature, 284, 135–9.

Ramseyer, K., J. Baumann, A. Matter, and J. Mullis, 1988. Cathodoluminescence colours of α-quartz. Mineralogical Magazine, 52, 669–77.

Ramseyer, K., J. Fischer, A. Matter, P. Eberhardt, and J. Geiss, 1989. A cathodoluminescence microscope for low intensity luminescence. Journal of Sedimentary Petrology, 59, 619–22.

Redmond, G., S. Kimoto, and H. Okuzumi, 1970. Use of the SEM in cathodoluminescence observations on natural samples. In Johari, O. (ed.), Scanning Electron Microscopy, Proceedings of the Third Annual Scanning Electron Microscope Symposium, Chicago, IL, IIT Research Institute, pp. 33–40.

Redmond, G., F. Cesbron, R. Chapoulie, *et al.*, 1992. Cathodoluminescence applied to the microcharacterization of mineral materials: a present status in experimentation and interpretation. Scanning Microscopy, 6, 23–68.

Redmond, G., M. R. Phillips, and C. Roques-Carmes, 2000. Importance of instrumental and experimental factors on the interpretation of cathodoluminescence data from wide band gap materials. In Pagel, M., V. Barbin, P. Blanc, and D. Ohnenstetter (eds.), Cathodoluminescence in Geosciences, Berlin, Springer-Verlag, pp. 60–126.

Reed, R. M. and K. L. Milliken, 2003. How to overcome imaging problems associated with carbonate minerals on SEM-based cathodoluminescence systems. Journal of Sedimentary Research, 73, 328–32.

Reed, S. J. B., 1995. Electron probe microanalysis. In Potts, P. J., J. F. W. Bowles, S. J. B. Reed, and M. R. Cave (eds.), Microprobe Techniques in the Earth Sciences, London, Chapman & Hall, pp. 49–89.

Reed, S. J. B. and I. M. Romanenko, 1995. Electron probe microanalysis. In Advanced mineralogy. Methods and Instrumentations: Results and Recent Developments, Berlin, Springer-Verlag, vol. 2, pp. 240–6.

Reeder, R. J., 1991. An overview of zoning in carbonate minerals. In *Luminescence Microscopy and Spectroscopy: Qualitative and Quantitative Applications*, SEPM Short Course 25, pp. 77–82.

Reinhold, C., 1998. Multiple episodes of dolomitization and dolomite recrystallization during shallow burial in Upper Jurassic shelf carbonates: eastern Swabian Alb, southern Germany. Sedimentary Geology, 121, p. 71–95.

Richter, D. K., Th. Götte, and D. Habermann, 2002. Cathodoluminescence of authigenic albite. Sedimentary Geology, 150, p. 367–74.

Richter, D. K., Th. Götte, J. Götze, and R. D. Neuser, 2003. Progress in application of cathodoluminescence (CL) in sedimentary petrology. Mineralogy and Petrology, 79, 127–66.

Richter, D. K. and U. Zinkernagel, 1981. Zur Anwendung der Kathodolumineszenz in der Karbonatpetrographie. Geologische Rundschau, 70, 1276–302.

Ridley, W. I. and F. E. Lichte, 1998. Major, trace, and ultratrace element analysis by laser ablation ICP-MS. In *Applications of Microanalytical Techniques to Understanding Mineralizing Processes*, Reviews in Economic Geology, 7, pp. 199–215.

Roedder, E., 1984. Fluid Inclusions: *an Introduction to Studies of all Types of Fluid Inclusions, gas, liquid, or Melt Trapped in Materials from Earth and Space, and their Application to the Understanding of Geologic Processes.* Mineralogical Society of America Reviews in Mineralogy, 12.

Ross, G. M., M. E. Villeneuve, and R. J. Theriault, 2001. Isotopic provenance of the lower Muskwa assemblage (Mesoproterozoic, Rocky Mountains, British Columbia): new clues to correlation and source areas. Precambrian Research, 111, 57–77.

Rowan, E. L., 1986. Cathodoluminescence zoning in hydrothermal dolomite cements: relationship to Mississippi Valley-type Pb–Zn mineralization in southern Missouri and northern Arkansas. In Hagni, R. D. (ed.), Process Mineralogy VI, Warrendale, PA, The Metallurgical Society Inc., pp. 69–87.

Rubatto, D. and D. Gebauer, 2000. Use of cathodoluminescence for U–Pb zircon dating by ion microprobe: some examples from the western Alps. In Pagel, M., V. Barbin, P. Blanc, and D. Ohnenstetter (eds.), Cathodoluminescence in Geosciences, Berlin, Springer-Verlag, pp. 373–400.

Rusk, B. and M. Reed, 2002. Scanning electron microscope–cathodoluminescence analysis of quartz reveals complex growth histories in veins from the Butte porphyry copper deposit, Montana. Geology, 30, 727–30.

Russ, J. C., 1984. Fundamentals of Energy Dispersive X-Ray Analysis, London, Butterworths, ch. 1.

Ryan, C. G., E. J. Clayton, W. L. Griffin, *et al.*, 1988. SNIP, a statistics–sensitive background treatment for the quantitative analysis of PIXE spectra in geoscience applications. Nuclear Instruments, and Methods in Physics Research B, 34, 396–402.

Sandberg, P. A., 1983. An oscillating trend in Phanerozoic non-skeletal carbonate mineralogy. Nature, 305, 19–22.

Schieber, J., D. Krinsley, and L. Riciputi, 2000. Diagenetic origin of quartz silt in mudstones and implications for silica cycling. Nature, 406, 981–5.

Scholle, P. A. and D. A. Ulmer-Scholle, 2003. A Color Guide to the Petrography of Carbonate Rocks: Grains, Textures, Porosity, Diagenesis, AAPG Memoir 77.

Schvoerer, M., P. Guibert, D. Piponnier, and F. Bechtel, 1986. Cathodoluminescence des mate?riaux arche?ologiques. PACT (Journal of the European Study Group on Physical, Chemical, Biological and Mathematical Techniques Applied to Archaeology), 15, 93–110.

Schweiger, A. and G. Jeschke, 2001. Principles of Pulse Electron Paramagnetic Resonance. Oxford, Oxford University Press.

Searl, A., 1988. The limitations of "cement stratigraphy" as revealed in some Lower Carboniferous oolites from south Wales. Sedimentary Geology, 57, 171–83.

Sedat, B., 1992. Petrographie und Diagenese von Sandsteinen im Nordwestdeutshen Oberkarbon. Hamburg, DGMK-Forschungsbericht 384–7.

Seyedolali, A., D. H. Krinsley, S. Boggs, Jr., *et al.*, 1997a. Provenance interpretation of quartz by scanning electron microscope–cathodoluminescence fabric analysis. Geology, 25, 787–90.

Seyedolali, A., S. Boggs, Jr., G. G. Goles, and D. H. Krinsley, 1997b. Cathodoluminescence of quartz from contact-metamorphosed rocks of

Skaergaard Intrusion and mechanically sheared metamorphosed rocks of Prescott, Arizona. Abstracts with Programs, Geological Society of America, 29(6), 401.

Sippel, R. F., 1965. Simple device for luminescence petrography. Review of Scientific Instruments, 36, 1556–8.

Sippel, R. F., 1968. Sandstone petrology, evidence from luminescence petrography. Journal of Sedimentary Petrology, 38, 530–54.

Smith, J. V. and R. C. Stenstrom, 1965. Electron-excited luminescence as a petrologic tool. Journal Of Geology, 73, 627–35.

Smith, K. C. A., 1956. The scanning electron microscope and its field of application. Ph.D. Dissertation, University of Cambridge.

Sprunt, E. S. and Nur, A., 1979. Microcracking and healing in granites: new evidence from cathodoluminescence. Science, 205, 495–7.

Stanley, S. M. and L. A. Hardie, 1999. Hypercalcification: paleontology links plate tectonics and geochemistry in sedimentology. GSA Today, 9, 1–7.

Steffens, P., E. Niehuis, T. Friese, D. Greifendorf, and A. Benninghoven, 1985. A time-of-flight mass spectrometer for static SIMS applications. Journal of Vacuum Science Technology A, 3(3), 1322–5.

Stenstrom, R. C. and J. V. Smith, 1964. Electron-excited luminescence as a petrologic tool. Geological Society of America Special Paper 76, p. 158.

Stevens Kalceff, M. A. and M. R. Phillips, 1995. Cathodoluminescence microcharacterization of the defect structure of quartz. Physical Review B, 52, 3122–34.

Stevens Kalceff, M. A., M. R. Phillips, A. R. Moon, and W. Kalceff, 2000. Cathodoluminescence microcharacterization of silicon dioxide polymorphs. In Pagel, M., V. Barbin, P. Blanc, and D. Ohnenstetter (eds.), Cathodoluminescence in Geosciences, Berlin, SpringerVerlag, pp. 193–224.

Stöffler, D. and F. Langenhorst, 1994. Shock metamorphism of quartz in nature and experiment: 1. Basic observations and theory. Meteoritics, 29, 155–81.

Stone, W. N. and R. Siever, 1996. Quantifying compaction, pressure solution and quartz cementation in moderately- and deeply-buried quartzose sandstones from the greater Green River Basin, Wyoming. In Siliciclastic Diagenesis and Fluid Flow, Society for Sedimentary Geology Special Publication 55, pp. 129–50.

Stow, D. A. V. and J. Miller, 1984. Mineralogy, petrology and diagenesis of sediments at Site 530, southeast Angola basin. In Hay, W. W., J. C. Sibuet *et al.* (eds.), Initial Reports of the Deep Sea Drilling Project, Washington, DC., US Government Printing Office, vol. LXXV, pp. 857–73.

Stünitz, H., 1998. Syndeformational recrystallization – dynamic or compositionally induced? Contributions to Mineralogy and Petrology, 131, 219–36.

Sylvester, P. (ed.), 2001. Laser-Ablation–ICPMS in the Earth Sciences: Principles and Applications. Ottawa, Mineralogical Association of Canada.

Tarashchan, A. N. and G. Waychunas, 1995. Interpretation of luminescence spectra in terms of band theory and crystal field theory. Sensitization and quenching, photoluminescence, radioluminescence, and cathodoluminescence. In Marfunmin, A. S. (ed.), Advanced Mineralogy 2, Methods and Instrumentations: Results and Recent Developments, Berlin, Springer-Verlag, pp. 124–35.

Taylor, J. M., 1950. Pore-space reduction in sandstones. American Association of Petroleum Geologists Bulletin, 34, 701–16.

Thornton, P. R., 1968. Scanning Electron Microscopy, London, Chapman and Hall.

Tobin, K. J., K. R. Walker, D. M. Steinhauff, and C. I. Mora, 1996. Fibrous calcite from the Ordovician of Tennessee: preservation of marine oxygen isotopic composition and its implications. Sedimentology, 43, 235–51.

Trewin, N., 1988, Use of the scanning electron microscope in sedimentology. In Tucker, M. (ed.), Techniques in Sedimentology, Oxford, Blackwell Scientific Publications, pp. 229–73.

van den Kerkhof, A. M. and U. F. Hein, 2001. Fluid inclusion petrography: Lithos, 55, 27–47.

van den Kerkhof, A. M., A. Kronz, and K. Simon, 2001. Trace element redistribution in metamorphic quartz and fluid inclusion modification: observations by cathodoluminescence. In Noronha, F., A. Do?ria, and A. Guedes (eds.), XVI ECROFI European Current Research on Fluid Inclusions, Porto 2001, Abstracts, Amsterdam, Elsevier, pp. 447–50.

Vortisch, W., D. Harding, and J. Morgan, 2003. Petrographic analysis using cathodoluminescence microscopy with simultaneous energy-dispersive X-ray spectroscopy. Mineralogy and Petrology, 79, 193–202.

Walderhaug, O. and J. Rykkje, 2000. Some examples of crystallographic orientation on the cathodoluminescence colors of quartz. Journal of Sedimentary Research, 70, 545–8.

Walkden, G. M. and J. R. Berry, 1984. Syntaxial overgrowths in muddy crinoidal limestones; cathodoluminescence sheds new light on an old problem. Sedimentology, 31, 251–67.

Walker, G., 2000. Physical parameters for the identification of luminescence centres in minerals. In Pagel, M., V. Barbin, P. Blanc, and D. Ohnenstetter (eds.) Cathodoluminescence in Geosciences, Berlin, Springer Verlag, pp. 23–39.

Walker, G. and S. Burley, 1991. Luminescence petrography and spectroscopic studies of diagenetic minerals. In *Luminescence Microscopy and Spectroscopy: Quantitative and Qualitative Applications*, SEPM Short Course 25, pp. 83–96.

Wallace, M. W., C. Kerans, P. W. Playford, and D. McManus, 1991. Burial diagenesis in the Upper Devonian reef complexes of the Geikie Gorge region, Canning Basin, Western Australia. American Association of Petroleum Geologists Bulletin, 75, 1018–38.

Watt, G. R., P. Wright, S. Galloway, and C. McLean, 1997. Cathodoluminescence and trace element zoning in quartz phenocrysts and xenocrysts. Geochimica et cosmochimica acta, 61, 4337–48.

Waychunas, G. A., 2002. Apatite luminescence. In Kohn, M. J., J. Rakovan, and J. M. Hughes (eds.), Reviews in Mineralogy and Geochemistry, Washington, DC, Mineralogical Society of America, vol. 48, 701–42.

Weil, J. A., 1984. A review of electron spin spectroscopy and its application to the study of paramagnetic defects in crystalline quartz. Physics and Chemistry of Minerals, 10, 149–65.

Weil, J. A., 1993. A review of the EPR spectroscopy of the point defects in α-quartz: the decade 1982–1992. In Helms, C. R. and B. E. Deal (eds.), The Physics and Chemistry of SiO_2 *and the Si-SiO₂ Interface 2*, New York, NY, Plenum Press, pp. 131–44.

Weil, J. A., J. R. Bolton, and J. E. Wertz, 1994. Electron Paramagnetic Resonance: Elementary Theory and Practical Applications, New York, NY, John Wiley & Sons, Inc.

Weil, J. A., Y. Dusausoy, and S. L. Votyakov, 1995. Electron paramagnetic resonance (EPR), 1995. In Marfunin, A. S., (ed.), Advanced Mineralogy, Methods and Instrumentations: Results and Recent Developments, Berlin, Springer-Verlag, vol. 2, pp. 197–209.

Williams, P. M. and A. D. Yoffe, 1968. Scanning electron microscope studies of cathodoluminescence in ZnSe single crystals. Philosophical Magazine, 18, 555–60.

Wilson, R. G., F. A. Stevie, and C. W. Magee, 1990. Secondary Ion Mass Spectrometry: A Practical Handbook for Depth Profiling and Bulk Impurity Analysis, New York, NY, John Wiley & Sons.

Wolfe, J. P., 1998. Imaging Phonons: Acoustic Wave Propagation in Solids, New York, NY, Cambridge University Press.

Worden, R. H. and S. D. Burley, 2003. Sandstone diagenesis: the evolution of sand to stone. In Burley, S. D. and R. H. Worden (eds.), Sandstone Diagenesis: Recent and Ancient, Malder, MA, Blackwell Publishing, pp. 3–44.

Wordon, R. H. and S. Morad, 2000. Quartz Cementation in Sandstones, International Association of Sedimentologists Special Publication, 29.

Yoo, C. M., J. M. Gregg, and K. L. Shelton, 2000. Dolomitization and dolomite neomorphism; Trenton and Black River limestones (Middle Ordovician) northern Indiana, USA. Journal of Sedimentary Research, 70, 265–74.

Young, S. W., 1976. Petrographic textures of detrital polycrystalline quartz as an aid to interpreting crystalline source rocks. Journal of Sedimentary Petrology, 46, 595–603.

Zinkernagel, U., 1978. Cathodoluminescence of quartz and its application to sandstone petrology. Contributions to Sedimentology, 8, 1–69.

Zuffa, G. G. (ed.), 1985. Provenance of Arenites, Dordrecht, D. Reidel Publishing.

Index